Soft Actuators and Robotics

Soft Actuators and Robotics

Editor

Hamed Rahimi Nohooji

Basel • Beijing • Wuhan • Barcelona • Belgrade • Novi Sad • Cluj • Manchester

Editor
Hamed Rahimi Nohooji
University of Luxembourg
Luxembourg

Editorial Office
MDPI
St. Alban-Anlage 66
4052 Basel, Switzerland

This is a reprint of articles from the Special Issue published online in the open access journal *Actuators* (ISSN 2076-0825) (available at: https://www.mdpi.com/journal/actuators/special_issues/NV7OH3RF9J).

For citation purposes, cite each article independently as indicated on the article page online and as indicated below:

Lastname, A.A.; Lastname, B.B. Article Title. *Journal Name* **Year**, *Volume Number*, Page Range.

ISBN 978-3-0365-9808-6 (Hbk)
ISBN 978-3-0365-9809-3 (PDF)
doi.org/10.3390/books978-3-0365-9809-3

Cover image courtesy of Hamed Rahimi Nohooji

Contents

About the Editor

Hamed Rahimi Nohooji

Dr. Hamed Rahimi Nohooji is a Research Associate at the University of Luxembourg, with a focus on soft robotics and human-robot interaction. His academic journey, crowned with a Ph.D. from Curtin University, has led him to work at renowned universities globally. His primary research drive is advancing control systems to improve human–robot synergy, a theme central to his role in scholarly publications within the robotics community.

Review

A Review of Soft Actuator Motion: Actuation, Design, Manufacturing and Applications

Xianzhi Tang [1], Huaqiang Li [1], Teng Ma [1], Yang Yang [2], Ji Luo [1], Haidan Wang [3] and Pei Jiang [1,*]

1 State Key Laboratory of Mechanical Transmission, University of Chongqing, Chongqing 400030, China
2 School of Automation, Nanjing University of Information Science and Technology, Nanjing 210044, China
3 Beijing Machinery Industry Automation Research Institute Co., Ltd., Beijing 100120, China
* Correspondence: peijiang@cqu.edu.cn

Abstract: Compared with traditional rigid robots, soft robots have high flexibility, low stiffness, and adaptability to unstructured environments, and as such have great application potential in scenarios such as fragile object grasping and human machine interaction. Similar to biological muscles, the soft actuator is one of the most important parts in soft robots, and can be activated by fluid, thermal, electricity, magnet, light, humidity, and chemical reaction. In this paper, existing principles and methods for actuation are reviewed. We summarize the preprogrammed and reprogrammed structures under different stimuli to achieve motions such as bending, linear, torsional, spiral. and composite motions, which could provide a guideline for new soft actuator designs. In addition, predominant manufacturing methods and application fields are introduced, and the challenges and future directions of soft actuators are discussed.

Keywords: soft actuator; actuation; design; manufacturing; applications

Citation: Tang, X.; Li, H.; Ma, T.; Yang, Y.; Luo, J.; Wang, H.; Jiang, P. A Review of Soft Actuator Motion: Actuation, Design, Manufacturing and Applications. *Actuators* **2022**, *11*, 331. https://doi.org/10.3390/act11110331

Academic Editor: Hamed Rahimi Nohooji

Received: 18 October 2022
Accepted: 11 November 2022
Published: 14 November 2022

1. Introduction

Traditional rigid robots are assembled from parts and structures with certain hardness and stiffness which are endowed with large load capacity and precise locomotion. However, it is difficult for these robots to conduct tasks in unstructured environments due to their high stiffness. Soft robots, show great advantages in flexibility, compliance, and adaptivity thanks to their soft materials and structures. They are widely adopted for grasping irregular objects [1–3] and minimally invasive surgery [4–6], which could broaden the working field of robots.

Soft actuators are crucial components of soft robots, and are fabricated by soft materials such as dielectric elastomers (DEs), electroactive polymers (EAPs), shape memory alloys/polymers (SMAs/SMPs), electroactive ceramics, thermoplastic polyurethane elastomers (TPUs), and hydrogels [7]. Soft actuators can be actuated by various stimuli, including fluid, thermal, electric, magnetic, light, humidity, chemical actuation, etc. Different actuation methods possess specific advantages. For instance, fluid actuation has the advantages of high energy efficiency, simple structure, flexibility and low cost [8–13]. Thermal actuation can avoid the need for bulky pumps and valve control systems to achieve wireless actuation [14–18]. Electric actuation has shown significant advantages in terms of fast response, non-loading stimulation manner, and easy integration [19–23]. Magnetic actuation can realize remote actuation for wireless soft actuators [24–27]. Light actuation has the advantages of remote directional and contactless actuation [28–30]. Humidity actuation can be realized by changing the humidity gradient of the environment [31–34]. Chemical actuation can be actuated by internal chemical reactions rather than external environmental stimuli [35]. Acoustic actuation is another potential actuation, which could transfer energy through sound waves to achieve locomotion [36–38].

Based on these actuation methods, specific motions are preprogrammed into the structures of actuators, which provide inexpensive and convenient methods for desired motions.

To achieve inchworm multimodal locomotion, Gu et al. presented a soft actuator that could complete a variety of locomotions, such as crawling, climbing, and transitioning between horizontal and vertical planes through the preprogrammed motion of the actuator [39]. To achieve object grasping purpose, Yoshida et al. proposed an internal exoskeleton gripper with variable stiffness pneumatic bending actuators based on low-melting-point-alloys, which was able to conduct a bending motion at different points and keep its bending shape without power input [40]. However, actuators with preprogrammed structures cannot achieve multiple motions by a single customized actuator. According to the preprogrammed structure, the constraints are embedded into the structure of the actuator, which cannot be changed to achieve specific motions. The reprogrammed structure of a soft actuator can adjust the structure constraints dynamically to achieve different deformations. Yoshida et al. presented an actuator in [41] which was able to achieve different motions (e.g., torsional and linear) by changing the fiber alignment. Yang et al. proposed a novel reprogrammable soft actuator which achieves spiral motion by tension jamming; the proposed jamming technology enables adaptive ability in the emerging actuator to deal with unknown environments [42]. Moreover, Cui et al. presented a soft microrobot which can be controlled to realize a specific shape transformation by programming magnetic configurations [43].

The presented literature provides a critical overview of the developments in soft actuators, including the actuation methods, preprogrammed structure design, and existing manufacturing types and applications, as shown in Figure 1. This article demonstrates the relations between actuation methods, preprogrammed structure design, and corresponding motions, which could provide a guideline for actuator design. At the end of this review, the continuing challenges and future directions of soft actuators are discussed.

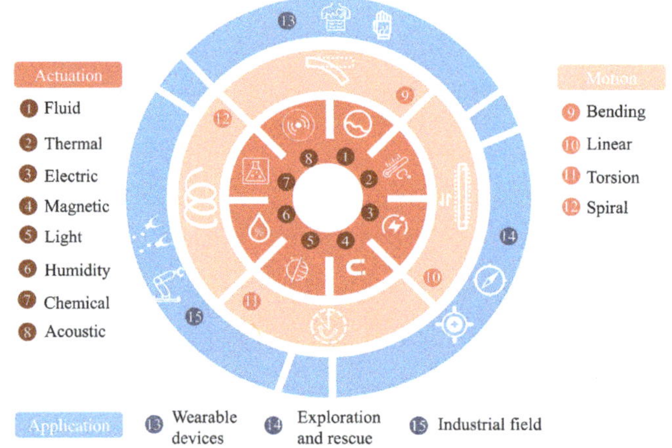

Figure 1. Actuation methods, motions, and applications of soft actuators.

2. Actuation

2.1. Fluid Actuation

Fluid actuation is one of the most common actuation methods for soft actuators, with high energy efficiency, large output force, a simple mechanism, and low cost. The medium of the actuation is divided into a compressible gas and incompressible liquid, which can be used to change the pressure of the sealed space made by flexible inflatable materials [44–47]. With changing volume in the confined space, expansion or contraction can be achieved. Moreover, the stiffness of flexible materials is the most important factor for this kind of actuation, as expansion or contraction always occurs first where the stiffness is minimal;

many motions can be achieved in this way. However, this actuation method is rarely used in wireless actuation, as it is usually accompanied by heavy auxiliary equipment.

2.2. Thermal Actuation

Thermal actuation is frequently used on intelligent materials which experience performance changes under application of thermal energy [48–50]. These materials can be divided into two different types: those in which the crystal structure of the material changes directly in the solid state, such as SMAs, and those in which the stiffness of the material is varied, such as SMPs.

Currently, SMAs are most widely used due to their high recovery stress. The crystal structure variation of nickel titanium (Ni-Ti) SMA is shown in Figure 2a. Specifically, the SMA can change modulus when it varies from austenite to martensite, which shrinks or expands from its initial shape with temperature variation [51–53].

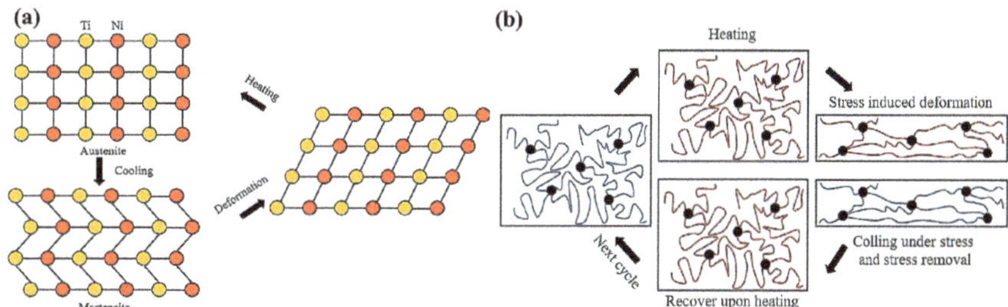

Figure 2. (**a**) The crystal structure variation of nickel titanium (Ni-Ti) SMA. (**b**) Schematic diagram of SMP deformation.

In addition, the stiffness ratio of SMPs can be as high as several hundred due to the change in the SMP molecular chain segments, as shown in Figure 2b. Specifically, the molecular chain segments transform from high stiffness to low stiffness when experiencing the transition temperature. The transition characteristics of SMPs provide an opportunity for realizing obvious deformations under the action of external force [54–58]. Wireless actuation can be achieved by tuning the temperature around the actuator. However, a significant drawback of this method of actuation is poor real-time ability and small output force.

2.3. Electric Actuation

Electric actuation has two main forms with respect to realizing functions. One is to change the stiffness of intelligent materials, such as electro-rheological fluids (ERF). The other is to make intelligent materials such as DE, ionic polymer–metal composites (IPMC), and EAPs deform directly.

The viscosity of ERF changes under the action of an electric field, which affects its stiffness [59,60]. Figure 3a shows the variation of ERF microstructure under electric actuation. ERF consists of dielectric particles and an insulating liquid, such as a Newtonian fluid. The randomly distributed dielectric particles are arranged in chains along the direction of the electric field. In this process, ERF changes from a high-viscosity gel state to a low-viscosity liquid state with stiffness variation.

Compared to ERF, DE can produce motion immediately under the action of direct current [61,62]. DE is squeezed by electrostatic force to deform with variable length when voltage is applied; the schematic diagram is shown in Figure 3b. Moreover, the Maxwell force is the electrostatic force which is generated between two electrodes under electrification. Due to this, the Maxwell force can achieve motion immediately with electric actuation [63]. However, DE and the Maxwell force both require high voltages to realize function. IPMC is an ionic electroactive polymer which is able to achieve deformation directly with a low actuation voltage based on the energy transfer mechanism of ion migration and its particular structure [64], as shown in Figure 3c. Specifically, an IPMC film consists of a positive electrode, a negative electrode, and an electrolyte. When a high voltage is applied to the electrodes, hydrophilic cations in the electrolyte accumulate at the negative electrode. Based on this, the IPMC film bends towards the positive electrode. The angle of bending is inversely proportional to the thickness of the IPMC film. Moreover, the geometrical parameters of the IPMC film, such as length, width and height, affect the deformation effect. IPMC does have disadvantages, hpwever, such as back relaxation effects. Electric actuation has the advantages of high response and remote control, although there exist problems with electrical breakdown coupling instability and poor safety.

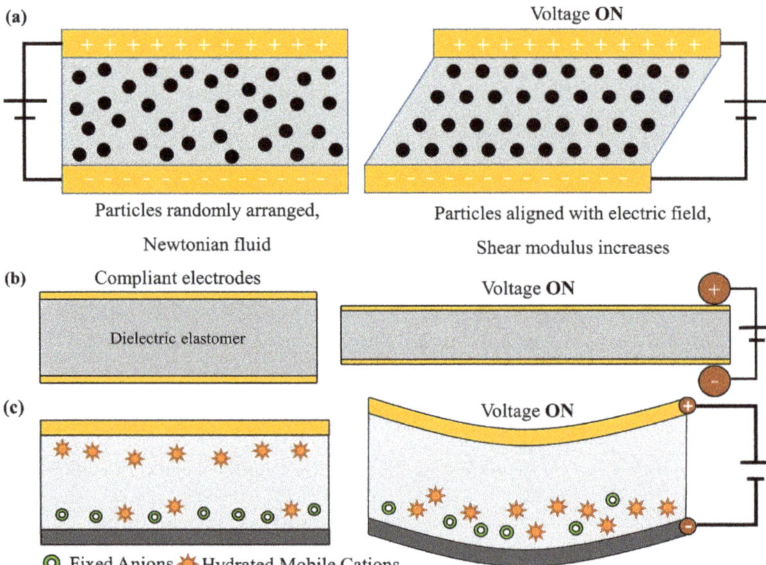

Figure 3. (**a**) Internal dielectric particle transformation of ERF after applying electric field. (**b**) Deformation schematic diagram of DE under voltage. (**c**) Deformation schematic diagram of IPMC under voltage.

2.4. Magnetic Actuation

Magnetic actuation plays an important role in adjusting the stiffness of magnetic-sensitive materials which can achieve deformation under external force by adjusting the magnetic field. Magneto-rheological fluid (MRF) is a magnetic sensitive material frequently used in this actuation; it consists of small soft magnetic particles with high permeability and a non-permeable liquid [65]. The principle of MRF stiffness modulation is based on the viscosity and yield stress of such fluids, which varies with changes in the applied magnetic field. As shown in Figure 4a, randomly distributed magnetic particles in the non-permeable liquid form particle chains along the direction of magnetic field. These particle chains can generate considerable viscosity and increase stiffness.

Compared with MRF, magneto-rheological elastomers (MRE) has better variable stiffness performance; the elastomer does not leak like a liquid does. The microstructure variation of MRE is shown in Figure 4b. MRE can be obtained by simply replacing the liquid medium with elastomers. The magnetic particles in MRE arrange into chains by magnetic fields during the curing process of the elastomers, similar to MRF [66].

Magnetic actuation can conduct manipulation with wireless actuation. However, it is difficult to control motions and force output, and it is hard to achieve precise control with the magnetic field.

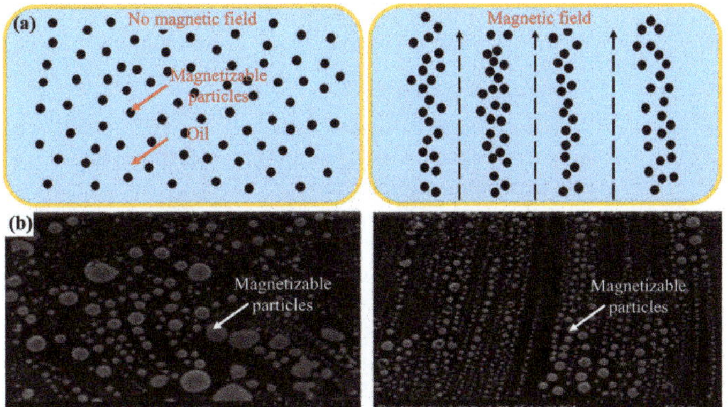

Figure 4. (**a**) Microstructure diagram of MRF. (**b**) Scanning electron microscopy photos of MRE microstructure variation.

2.5. Light Actuation

Light actuation uses an external stimulus to deform the materials via exposure to light. There exist two primary forms of light actuation, namely, photothermal and photochemical actuation, which use different mechanisms. The former is actuated by light-induced temperature transitions, while in the latter photosensitive materials are directly actuated to realize deformation.

Photothermal actuation converts the absorbed light energy into heat energy through the photothermal material in the system; the thermal energy induces the deformation of the polymer material. The photoinduced shape memory polymer material is prepared by introducing photothermal materials into the polymer matrix. This kind of material is essentially a thermally-induced shape memory material. The optical fiber is introduced into the polystyrene matrix to construct a photoinduced shape memory polymer. After the material is shaped, it is irradiated with a mid-infrared laser [67]. The optical fiber converts the light energy into heat energy and induces the material to return to its initial shape. Figure 5a shows the shape recovery of the material at different irradiation times.

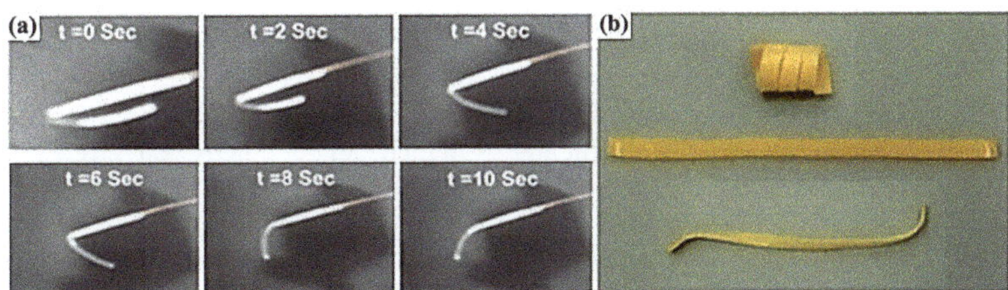

Figure 5. (**a**) Shape recovery of the photoinduced shape memory polymer material [67]. (**b**) Deformation process of photothermal shape memory polymer materials [68].

In addition, photochemical actuation is actuated by a chemical reaction that occurs after illumination. The groups with photochemical reactivity are introduced into the polymer network to construct memory materials, and the materials deform and restore the initial shape via light-controlled chemical changes. For instance, when light is irradiated on one side of the material, the monomer molecules on the irradiated side become crosslinked. This contraction only occurs on the irradiated side, while the non-irradiated side remains unchanged, causing the entire material to undergo deformation [68], as shown in Figure 5b.

Moreover, light actuation can achieve fixed-point actuation by using a small range of light with infinite energy sources. Based on this, wireless actution could be realized without the constraint of connectors. However, there exist problems with slow reaction times, small output force, and poor controllability.

2.6. Humidity Actuation

Humidity actuation is based on the characteristic of moisture-sensitive materials, which can realize internal volume changes by adjusting the environmental humidity gradient to accomplish deformation.

Polymer films are commonly used as moisture-sensitive materials; they achieve deformation by absorbing and releasing water, allowing them to quickly respond to environmental humidity changes and realize bending deformation under the effect of water vapor [69]. A schematic diagram is shown in Figure 6a. Cellulose is able to synthesize cellulose stearates with different degrees of substitution. Cellulose stearates with a low degree of substitution have good humidity response characteristics. Water molecules can be absorbed or released by the cellulose stearate film. Hence, the cellulose stearates can achieve controllable bending and folding movement [70], as shown in Figure 6b.

However, humidity actuation is usually accompanied by relatively difficult control and small output force. It has challenges with accomplishing certain operations, such as load bearing, and is unable to achieve long-term stable actuation performance.

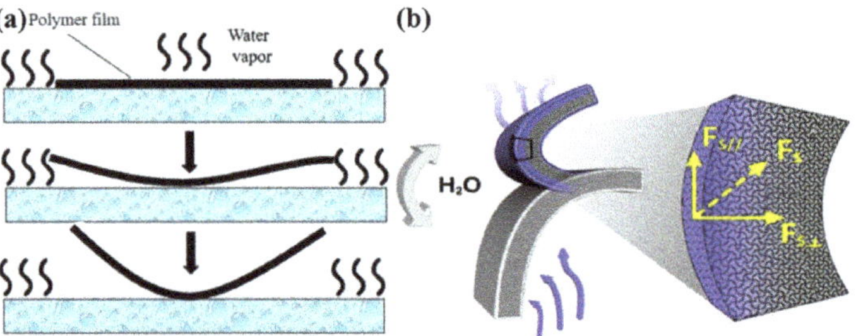

Figure 6. (**a**) Schematic diagram of polymer film with water vapor. (**b**) Schematic representation of cellulose stearate film with the effect of water.

2.7. Chemical Actuation

Chemical actuation is different from other actuations that are actuated by external stimuli. The actuation power comes from the energy generated by combustion or chemical reactions, which is used to achieve motion.

Combustion reactions exhibit dramatic changes and quickly produce a large amount of gas and heat energy, which can be directly converted into mechanical energy for actuation [71]. However, combustion reactions make it difficult to accurately control the resulting motions and improve the robustness due to the multiple factors of combustion. Moreover, chemical reactions may produce gas slowly, and these gases can be accumulated as a gas source to conduct motions [72]. For instance, liquid solvents under catalysis can create a large amount of gas. The chemical reaction of hydrogen peroxide catalyzed by silver produces oxygen; the schematic diagram of this chemical reaction is shown in Figure 7. Chemical actuation is an internal actuation method that can achieve wireless actuation. However, it cannot be active for very long without external energy input, and it is difficult to control the actuation energy.

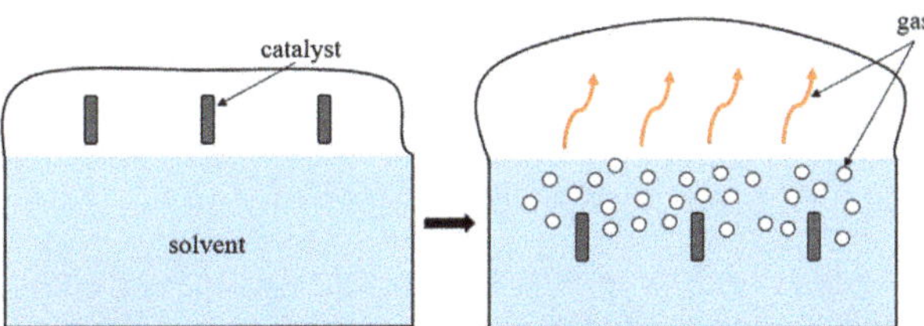

Figure 7. Schematic diagram of the reaction between hydrogen peroxide and catalytic silver.

2.8. Acoustic Actuation

Acoustic actuation provides a new opportunity to implement wireless actuation. Theoretically, acoustic energy could suspend objects in the air or manipulate objects. Hopefully, this potential method can extend the choices available for actuation methods. The principle of manipulation is to use standing waves to form sound traps, which then create suspension forces to resist gravity and keep items suspended [73], as shown in Figure 8a.

Furthermore, acoustic actuation may become a new actuation method through the use of ultrasonics to actuate particular liquids, such as non-Newtonian liquids. When actuated by ultrasonic waves, a non-Newtonian liquid is pushed up and deformed, as shown in Figure 8b. If elasticity or film could be used to limit the deformation of non-newtonian liquid, the actuation principle could be used to actuate a soft actuator in a similar manner to fluid actuation [74].

Table 1 summarizes the advantages and disadvantages of each actuation method. In order to achieve a certain operation or motion, it is important to select the appropriate actuation method to provide actuation energy based on their respective advantages. Furthermore, combining multiple actuation methods may make it possible to overcome the limitations and disadvantages of a single method.

Table 1. Advantages and disadvantages of various actuation methods.

Actuation	Advangtages	Disadvantages	References
Fliud actuation	high energy efficiency, simple structure, flexibility, low cost.	heavy auxiliary equipment, high sealing requirements.	[44–47]
Thermal actuation	contactless actuation, miniaturized actuation.	poor controllability, low actuation efficiency, dependence of intelligent materials.	[48–58]
Electric actuation	high response, wireless actuation, easy integration.	coupling instability of electric breakdown, and poor safty.	[59–64]
Magnetic actuation	remote control without contact.	poor precision control	[65,66]
Light actuation	remote directional wireless actuation.	slow reaction, small output force, poor controllability.	[67,68]
Humidity actuation	wireless actuation.	low precision deformation, poor real-time ability.	[69,70]
Chemical actuation	internal chemical reactions, rapid response.	low precision, poor controllability.	[71,72]
Acoustic actuation	non-contact actuation, direct energy conversion.	small actuation force, complex auxiliary equipments.	[73,74]

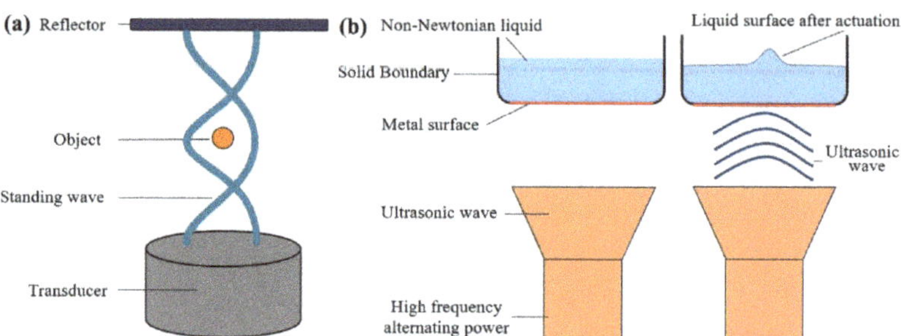

Figure 8. (**a**) Schematic of an object suspended by a standing wave. (**b**) Schematic diagram of a non-Newtonian liquid actuated by an ultrasonic wave.

3. Design

3.1. Bending Motion

By adding non-expansive constraints to flexible materials, bending motions can be achieved under gas pressure. Non-expansive constraints can be preprogrammed at a specific position of the soft actuator to constrain the deformation of the actuator, which can result in uneven stretching to achieve bending motion. An actuator based on a fiber-reinforced structure was proposed in [75] which consists of a soft gas chamber, circumferential winding fibers, and a strain-limiting layer, shown in Figure 9a. The double helix winding fibers around the gas chamber constrain the radial expansion, and the strain-limiting layer at the bottom of the gas chamber limits the elongation of one side of the actuator. The preprogrammed fiber-reinforced structure results in a bending motion towards the strain-limiting layer when inflated. In addition, the angle of bending and the input pressure vary approximately linearly [75]. Compared with a multi-chamber pneumatic bending actuator, this bending actuator had a larger output force, and the bending curvature was constant. Galloway et al. presented a fiber-reinforced soft bending actuator. The bending curvature and axis of the actuator can be programed with a flexible and selective placement of conformal coverings [76]. Ye et al. designed a pneumatic soft bending gripper; the bending motion is generated by using a limiting fiber to replace the axial restraint layer [77]. Deimel et al. proposed a highly compliant bending soft actuator, with the semicircular chamber inside the actuator is fiber-reinforced as well. Bending motion can be achieved due to the decreasing cross-section along the root to the tip [78].

Uneven stretching of a soft structure can be used to achieve a bending motion when pressurized. The expansion of the actuator in the axial direction causes the actuator to elongate unevenly under pneumatic actuation, which leads the actuator to bend in a less elongated direction. A pneumatic networks bending actuator was proposed by Mosadegh et al. [**?**], as shown in Figure 9b. The actuator is based on multi-chamber structure with elastic material, and the parallel chambers are connected by a channel. The inextensible bottom layer acts as a limiting layer, and the asymmetric elongation of chambers and limiting layer induce the multi-chamber to conduct a bending motion when pressurized. Additionally, Shepherd et al. designed a pneumatic network of flexible actuators that bend under inflation due to the decreasing cross-section of the chambers [80]. Based on a number of origami pneumatic networks chambers, Kim et al. proposed a deployable soft actuator. The sides of the chambers are combined with a non-elongable structure, while the axial direction can freely extend under pressure. When the uneven elongation of the side and the axial direction induce the actuator to bend, the origami pneumatic networks chambers rapidly expand and increase the force arm when inflated [81].

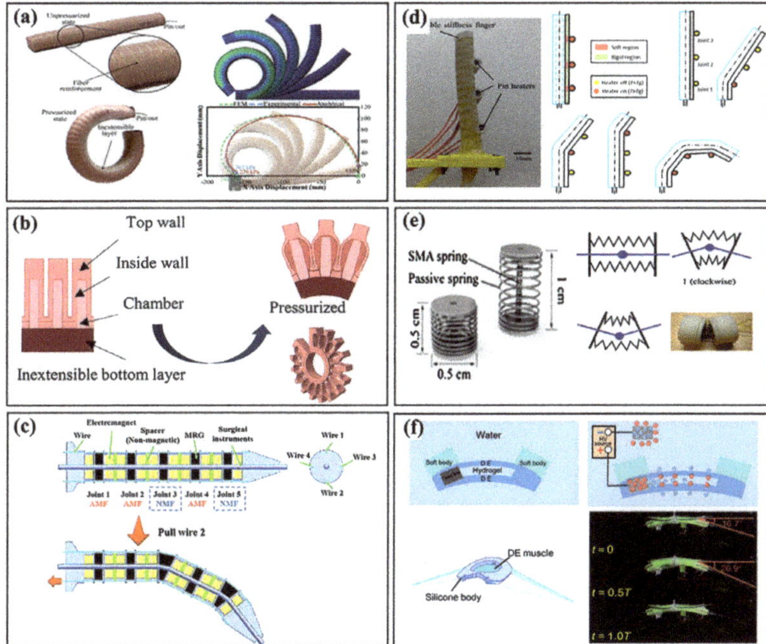

Figure 9. (**a**) The bending motion and finite element analysis of a fiber-reinforced soft actuator [75]. (**b**) Pneumatic network structure and bending posture under pressure [?]. (**c**) Bending motion of the variable stiffness multi-joint actuator [82]. (**d**) Actuator prototype with bending status of each joint [83]. (**e**) Single spring structure and bending posture combining two spring units [84]. (**f**) The internal structure and actuation principle of a bending actuator [85].

In addition, variable stiffness structures can achieve bending motion at the low stiffness point when actuated. By changing the structure stiffness of the actuator parts, the parts with low stiffness first deform under the actuation force. By reasonably designing and arranging the distribution of uneven stiffness, the actuator is able to achieve bending motion. Utilizing intelligent materials, variable-stiffness structures could be more widely adopted in soft actuators. Kitano et al. presented a soft actuator to accomplish bending motion by adjusting joint stiffness based on a magneto-rheological (MR) gel [82], as shown in Figure 9c. The actuator consists of MR gel rings, electromagnets, four uniformly distributed wires, and non-magnetic gaskets. The primary function of the gaskets is to prevent leakage of the magnetic field from affecting the performance of adjacent joints. When a magnetic field is applied to the MR gel ring, the stiffness increases rapidly due to the large variation in gel viscosity. On the contrary, when the magnetic field disappears, the stiffness of the MR gels decreases. Consequently, the MR gel rings become soft joints, which bend under the unbalanced force of the pulling wires. Similarly, Yang et al. designed a soft actuator that could deform in the same way as a human finger [83], as shown in Figure 9d. According to the design, a fiber-reinforced elastomer pastes to a substrate made of SMP. The heaters distribute at the bottom of the SMP substrate. According to the characteristics of SMP [86], the stiffness decreases when heated, and the bending motions happens around the heater when the elastomer is pressurized. In addition, Shintake et al. designed a variable stiffness bending actuator which is actuated by DE, with stiffness controlled by a low melting point alloy (LMPA) [49]. Zhang et al. designed a two-layer actuator which consists of a light-absorbing layer and an active layer [87]. The light-absorbing layer is made up of single-walled carbon nanotube; it absorbs light energy and converts it into heat energy. The active layer consists of a polymer that can expand or contract when heated.

Hence, the bending motion is achieve due to the uneven expansion or contraction of the two-layer structure.

Thus, actuators can accomplish bending motion based on unbalanced force. Unbalanced force generated by smart materials or structural design is distributed on different sides of the actuator by proper structural design, which allows the actuator to deform towards the side with greater force. For instance, a small soft actuator based on an SMA spring was designed by Yuk et al. [84], and is shown in Figure 9e. Two combined springs are distributed on both sides of the actuator. The SMA spring activated by electric current shrinks to 50% of the initial state. In this way, the unbalanced contracting force in the couple of springs induces a bending motion. Similarly, Li et al. designed a soft actuator made up of a pre-stretched dielectric elastomer (DE) and an ionically conductive hydrogel [85], two silicone films are attached to each end of the DE acting as wings, as shown in Figure 9f. The antagonism between the pre-stretched force and Maxwell force generated by the electric field leads to a bending motion. Hence, a bionic fish has been proposed based on this effect which is able to swim underwater similar to fish. However, many bending soft actuators remain in the laboratory stage, and are rarely used in specific task environments due to various limitations, such as insufficient force output and difficulty in achieving precise control.

3.2. Linear Motion

Soft actuators have been presented to achieve linear motion via circunferential or axial winding fibers. Fibers wound to the circumferential or axial surface of the actuator can constrain expansion in the corresponding direction of the actuator to produce stretching in a specific linear direction. McKibben artificial muscles were first proposed to achieve linear motion in the 1950s [88]. The typical McKibben artificial muscle consists of an inner soft tube and a double-helix braided sheath with nonextensible fibers, as shown in Figure 10a. When the inner tube is inflated, the high-pressure gas compresses the inner surface and the outer sheath, which has a tendency to increase its volume. Due to the non-scalability (or high longitudinal stiffness) of the fibers in the woven reticulated sheath, the actuator achieves inner tube elongation as its volume increases. There exists a linear relationship between the input pressure and the stretch length of the McKibben artificial muscles [88]. Homoplastically, Yamaguchi et al. proposed an in-pipe mobile actuator with two clamping units. One is limited axially by fiber and the other, used as the propelling unitm is wound circumferentially [89], as shown in Figure 10b. The propelling unit can elongate in the axial direction when inflated because the circumferential fiber limits the radial expansion. Combined with the clamping units, the propelling unit can realize continuous linear motion. This actuator was able to complete a crawling movement in a pipe by controlling these three units sequentially.

Certain intelligent materials can conduct linear expansion or constriction under external stimulated conditions. Thanks to the properties of smart materials, it is possible for them to stretch under heating or electrostatic forces, which can result in linear motion by combining multiple actuators to amplify tiny stretch deformations in each actuator. Menciassi et al. presented a linear motion soft actuator [90], which is shown in Figure 10c. The actuator consists of a flexible SMA spring, two brass disks, and a silicon shell. The two brass disks are combined in the silicon shell, which is connected to the SMA spring. Contraction of the SMA spring pulls the silicone shell when the Joule effect heats the spring. After the SMA spring cools, the silicon shell is restored to the initial state. Consequently, the actuator can conduct linear motion. The actuator was used as an earthworm to achieve linear peristalsis and bionic movement by asynchronous control. Ni et al. demonstrated a polymer interdigitated pillar electrostatic (PIPE) actuator based on dielectric liquid which achieved an output force density 5 to 10 times higher than natural muscle [91], as shown in Figure 10d. The PIPE actuator consists of chips, dielectric liquid, and springs; each chip has a high density column, while the spring between the chips connects the two parts and provides the restoring force. Notably, the dielectric liquid fills the interspace, which could increase breakdown voltage and electrostatic force. When an electric field is applied, the generated electrostatic force between the two chips pulls the movable pillar array towards the fixed pillar array without changing the pillar space, resulting in linear contraction. After the electric circuit is cut off, the springs are restored to their original position. Meanwhile, larger force output can be achieved by combining multiple PIPE actuators.

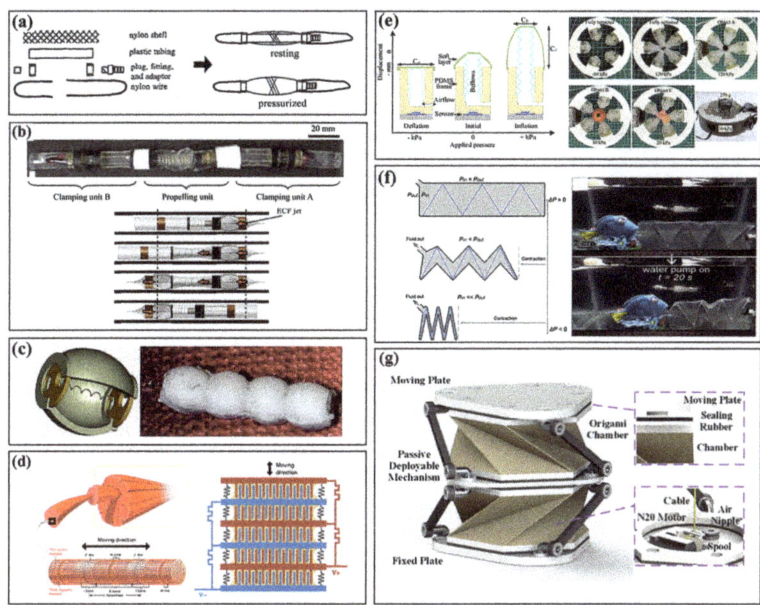

Figure 10. (**a**) The composition structure of McKibben artificial muscle and the forms in two states [88]. (**b**) The in-pipe mobile robot and locomotion of the in-pipe mobile robot [89]. (**c**) The structure of the module and the artificial earthworm prototype [90]. (**d**) Structure of skeletal muscles and stack schematic diagram of polymer interdigitated pillar electrostatic (PIPE) actuators, which are similar to the structure of natural muscles [91]. (**e**) Conceptual model of the bellows-driven soft actuator that represents smooth muscle contractions of a stomach [92]. (**f**) Schematic diagram of the operation of an actuator pulling an underwater fish for 3.5 cm in 20 s [93]. (**g**) Schematic diagram of a pneumatic/cable-driven hybrid linear actuator [94].

At present, many foldable structures are used to achieve linear motion with a symmetrical cross-section, which can produce a telescopic motion in a fixed direction through the design of the crease under the pneumatic or cable actuation. A bellows was used as a kind of foldable structure with a soft actuator to achieve linear motion in [92]. The actuator was able to produce linear motion similar to the smooth muscle in the stomach, as shown in Figure 10e. After gas is infalted or extracted, the top of the bellows is subjected to internal or external gas pressure. Based on this, the bellows can conduct linear motion. Due to circumferential creases, the radial direction has a resistance force that resists the gas pressure, and there is slight radial expansion or contraction; the main change is the axial expansion. Based on the bellows, a soft pneumatic actuator was made and installed in a circular frame to simulate the circular contraction of the stomach. Similarly, origami is one kind of foldable structure with light weight and high efficiency. Based on origami, Li et al. presented an origami-inspired artificial muscle which could realize linear motion like a fish [93], as shown in Figure 10f. The origami structure requires only a foldable skeleton, a flexible skin, and a fluid medium. The linear motion of contraction or elongation is realized by changing the pressure difference between the inside and outside of the artificial muscle. In the initial equilibrium state, the pressure difference between internal and external fluid is zero. When the pressure drops, the external skin produces tension, thus actuating the foldable skeleton of origami to achieve linear motion. Moreover, Zhang et al. designed a pneumatic/cable-driven hybrid linear actuator with a deployable mechanism based on an origami structure [94], as shown in Figure 10g. The structure mainly includes a pneumatic folding room, a cable drive system, and an unfolding mechanism. Pneumatic pressure makes the actuator elongate, and the pulling cable makes the actuator shorten. Bidirectional linear motion is realized through the resistance of the cable and pneumatic pressure. Because an origami chamber has a coupled torsional motion, two folding chambers with reverse torsional motion are used to offset the torsional motion to produce a pure linear flat motion. Notably, the linear actuator contains a passive deployable mechanism with high radial stiffness, which is able to withstand large loads. In addition, Aziz et al. designed a novel coil polymer actuator that could achieve linear motion by converting microwave radiation into thermal stimulation [95]. This spiral rotating actuator is made of high-stretch nylon-6 fibers filled with carbon nanotube (CNT). Under microwave irradiation, the CNTs absorb external electromagnetic energy and convert the energy into thermal energy to heat the fibers, allowing the actuator to achieve linear motion similar to muscle.

3.3. *Torsional Motion*

Non-uniform stiffness structures of actuators can be designed to achieve torsional motion. Torsional motion can be generated by means of preprogrammed designs, as non-uniform stiffness structures are able to exhibit different deformations under actuation. Gorissen et al. proposed a flexible pneumatic torsional actuator [96] which is shown in Figure 11a. The torsional actuator consists of two identical back-to-back pneumatic airbag arrays; the chambers of thepneumatic airbag are tilted and in a uniform arrangement. When the two pneumatic airbags are inflated separately, the actuator achieves bidirectional torsional motion by actuating two pneumatic airbag actuators separately. Moreover, Xiao et al. designed a torsional pneumatic networks actuator with three spiral chambers distributed on the elastomeric body, which collapse inward cooperatively when negative pressure acts on the chambers, resulting in torsional motion [97]. Analogously, a foldable actuator with non-uniform stiffness could accomplish torsional motion. Jiao et al. presented a foldable bidirectional torsional actuator with creases [98].The crease is obtained by removing part of the material on the surface of the soft actuator. As the thickness of the crease is smaller than the wall thickness, the stiffness at the crease is lower. When the vacuum is extracted, shrinkage first occurs at the crease, and this shrinkage can be used to achieve torsional motion.

An asymmetric actuation force can be generated due to asymmetric distribution of intelligent materials in the preprogrammed structure, resulting in different deformatoins used to achieve torsional motion. Shim et al. designed a torsional actuator composed of an SMA wire embedded in a polydimethylsiloxane (PDMS) matrix [99], as shown in Figure 11b. As the main mechanism of actuation, the SMA wire has torsional strain preapplied. The prestrained SMA wire shrinks during the martensite to austenite phase transformation under Joule heat. The SMA wire then returns to the initial state without torsional prestrain when shinking. Hence, the actuator achieves torsional motion with a large torsional angle. Additionally, when energized under clockwise torsional prestrain, the achieved effect is counterclockwise torsion. Ahnet al. proposed a soft morphing actuator using smart soft composite composed of SMA, PDMS, and acrylonitrile butadiene styrene (ABS) [100]. The actuator achieves torsional motion via the coupling effect of motions induced by the composite. However, these actuators are very small and produce minimal output force.

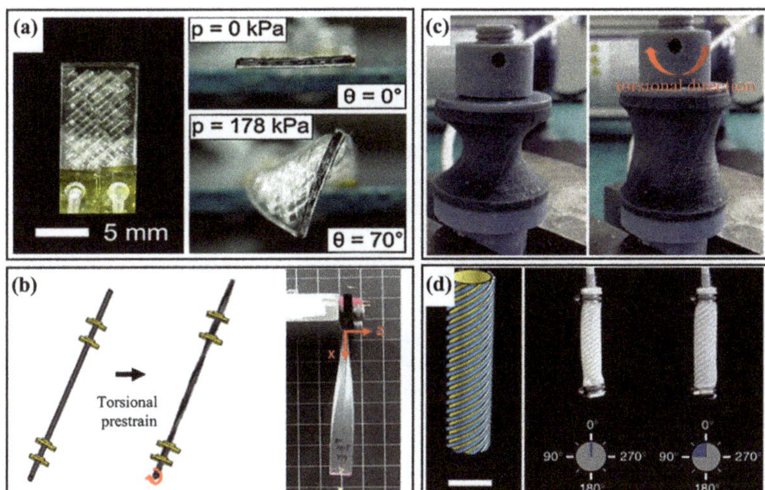

Figure 11. (**a**) Torsional actuator and torsional angles with different pressures [96]. (**b**) Schematic diagram of torsional prestrain and torsional angle of the actuator [99]. (**c**) Process of a pneumatic torsional actuator [101]. (**d**) Model drawing and torsional effect [102].

Asymmetric constraints can be imposed on pneumatic preprogrammed structures to conduct torsional motion. For example, asymmetric constraints with a spiral distribution can be subjected to a stretching force, causing the actuator to produce a torsional motion under the pneumatic actuation. These kinds of architectures can be used to achieve large overall motion and output force. Lee et al. proposed a new pneumatic modular torsional actuator with a spiral constraint added to its surface [101], as shown in Figure 11c. When applied to compressed gas, the actuator module induces torsional motion due to the spiral constraint. However, the torsional deformation of the actuator is not obvious due to the length limitation of the fibers embedded in the actuator. There exists a small angle when the pressure gas is supplied. To improve torsional performance and motion, a torsional actuator was designed by Schaffner et al. [102], as shown in Figure 11d. This actuator can achieve a large torsional angle under pneumatic actuation. Sprial stripes are added to the outer surface of the elastic body and fixed at both ends. When the elastic body chamber is inflated, a torsional motion occurs due to the constraint of the external spiral stripes. In addition, Yan et al. presented an inflatable soft actuator driven by two spiral chambers wound by fibers. This actuator has a purely efficient torsional motion, without any bending or extension movements [103].

3.4. Spiral Motion

Based on the non-uniform shrinkage of intelligent materials, spiral motion can be achieved as well. Non-uniform shrinkage is usually used in bilayer structures. The shrinkage of one layer is usually greater than that of the other layer. When the degree of shrinkage varies greatly, the bilayer structure achieves a spiral motion. Feng et al. designed a bilayer structure to accomplish spiral motion [104], whoich is shown in Figure 12a. The bilayer structure is 3D printed and corresponding self-morphing wire material is polylactic acid (PLA); the stresses of the two layers are in different directions due to different printing patterns. When the bilayer structure is heated to above the glass transition temperature, the upper layer shrinks and the lower layer expands along the printing angle, which leads to self-spiralling behavior. Notably, the right screw and left screw can be controlled by the printing angle. However, this kind of motion is limited by small output force and fewer applications because of the material characteristics.

Furthermore, the coupling effect between different motions can be combined to realize sprial motion. For example, the combination of bending and torsional motion can achieve spiral motion. Hoang et al. designed a variable stiffness spiral gripper [105], shown in Figure 12b. The gripper is mainly composed of a core soft actuator of fluid actuation, fabric sleeves, and a variable stiffness structure. When pressurized fluid actuates the core actuator, a coupled extended torsional motion is generated. Meanwhile, the gripper finger causes a bending motion, with a strain limiting the side of the fabric sleeve. Hence, the gripper is able to accomplish a spiral motion to wind around objects by combining bending and torsional motion.

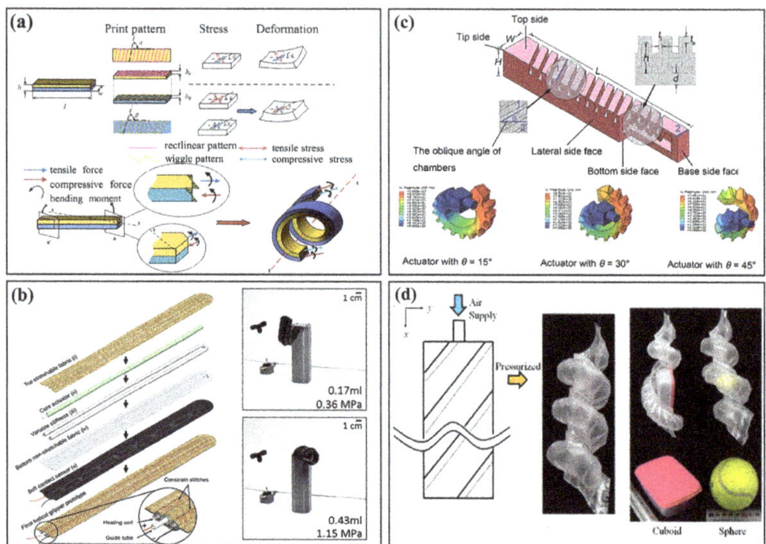

Figure 12. (**a**) Schematic diagram of a bilayer structure under different printing patterns and the principle of deformation [104]. (**b**) Structure of a spiral actuator and spiral deformation posture [105]. (**c**) Structure diagram of an actuator and spiral posture under different pressures [106]. (**d**) Structure diagram of actuator and spiral effect [107].

In addition, a preprogrammed asymmetric structure can be used to accomplish spiral motion. A tilted chamber design allows the actuator to elongate in the vertical direction of the tilted angle. Due to the existence of constraints, spiral motion can be achieved under pneumatic actuation. Wang et al. proposed a spiral gripper by designing asymmetric inclined chambers [106], as shown in Figure 12c. Chambers with a certain inclination angle expand via inflating the high-pressure gas; the expansion direction is perpendicular to the inclination angle of the cavity, which accomplishes spiral motion of the winding objects. Alternatively, soft actuators can realize spiral motion by using a plastic film to obtain better grasping force. Compared with an actuator made of elastic material, a folding plastic film can result in a lightweight soft actuator which is suitable for the human body to wear and has enough durability for use. Amase et al. presented the spiral gripper [107] shown in Figure 12d. The gripper consists of two layers of plastic film. The upper layer is folded obliquely to produce creases, then the two layers are connected by thermal welding. When inflated, the gripper realizes a spiral motion, and can automatically change its shape according to the shape of the object.

3.5. Composite Motions

The actuators above are only able to conduct a specific motion when actuated, which is not suitable for complex tasks. Therefore, actuators that can achieve composite motions attract more attention. For multi-module actuators, each module is able to achieve different motion, and composite motions can be realized by combining multiple modules. Due to pneumatic artificial muscles and external fiber constraints, Guan et al. designed a kind of actuator that could be assembled modularly [108], as shown in Figure 13a. Pneumatic artificial muscles achieve elongation and contraction, while bending and spiral motion are achieved by adding parallel axial and helical cables; then, by combining various bending and spiral actuators, the result is a combined actuator that can conduct complex composite motions, similar to an elephant's trunk.

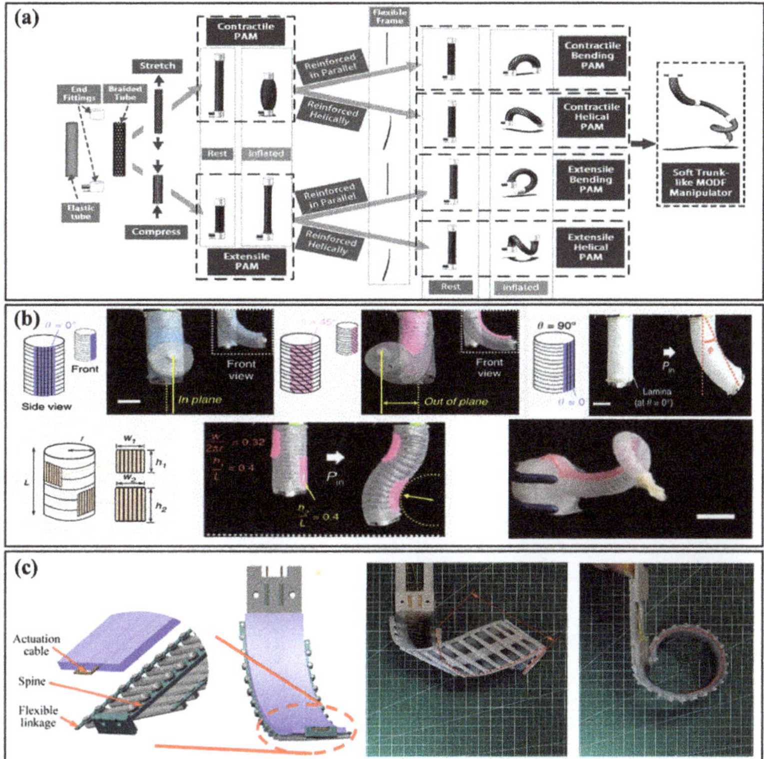

Figure 13. (**a**) The realization process of pneumatic artificial muscle composite motions [108]. (**b**) Motion effects of three angle fiber direction composite laminae adhered on a soft actuator [109]. (**c**) The mechanism of a scaffold reinforcement soft actuator and the effect of different motions by changing the scaffold direction [110].

More dexterous motions can be accomplished by dynamically adding external constraints. These dynamically added constraints can make the actuator achieve specific deformations, which can then be combined to achieve more complex motions. Kim et al. presented a self-adhesive composite laminate [109] which was able to adhere to any volume-expanding soft body to control its trajectory. Due to the unidirectional embedding of non-expandable continuous fibers in the hyperelastic matrix of composite laminates, these composite laminates can only be stretched in one direction. The soft actuators can deform specific trajectories according to their design by simply adhering laminates to their surfaces. Moreover, self-adhesive composite laminates have multiple fiber orientations to achieve bending or spiral movements by adhering, such as 0, 45°, and 90°fiber orientations. Figure 13b shows the various motions after adhering laminates on the surface of such a soft actuator; the composite laminates can adhere to any place in order to achieve motion.

These variable-constraint structures can be used to increase the precise control and motion diversity of organisms. Target motion effects with high application significance could be achieved by adjusting the constraints.

In addition, such reprogrammed structures could be used to realize different kinds of motions, which would then adjust the distribution of the constraints on the actuator and could achieve different deformations, such as bending and spiral motions. Recently, Jiang et al. proposed a soft gripper consisting of a soft layer and a rotatable scaffold reinforcement layer, where the scaffold reinforcement layer is distributed on the surface of the soft gripper [110], as shown in Figure 13c. The actuation cable is inside the soft layer, while the scaffold layer is made of a spine and flexible linkages. The direction of the scaffold can be adjusted by controlling the cable, which in turn can change the motion constraint of the gripper, resulting in different motion forms. By dynamically adjusting the direction of the scaffold, the bending motion and complex spiral motion of the soft gripper in three-dimensional space can be accomplished via the actuation cable.

A structure design matrix is proposed to summarize existing implementations of various motions through different actuation methods and design approaches, as shown in Table 2. To the best of our knowledge, actuators able to conduct certain motions under specific actuation have not yet been reported in literature. Designs of the corresponding actuators could refer to structures of actuators under similar actuation methods. For instance, actuators under chemical actuation have not been reported to achieve torsional motions. Because chemical actuation usually makes use of the gas generated in chemical reactions to pressurize actuators, the actuator structures under fluid actuation, such as tilted multi-chamber and external spiral constraints, could be referenced to design torsional actuators under chemical actuation.

Table 2. Structure design matrix based on actuations and motions.

Motion / Actuation	Bending	Linear	Torsional	Spiral
Fluid	Uneven constraints, Asymmetric stretching.	Fiber-reinforcement, Uniform constraints, Foldable structure.	Multi-chamber, Asymmetric constraints, Uneven elongation.	Tilted multi-chamber.
Thermal	Variable stiffness, Non-uniform stiffness.	SMA spring structure.	Torsional prestain.	Bilayer uneven shrinkage.
Electric	Variable stiffness structures, Unbalanced force.	Linear stretching materials.	/	/
Magnetic	Variable stiffness structures.	/	/	/
Light	Non-uniform shrinkage.	/	/	Non-uniform shrinkage.
Humidity	Internal small volume variation.	/	/	/
Chemical	Uneven volume expansion.	/	/	/

In addition, the materials used in soft actuators are an important part in the design and manufacture of actuators. Different materials have different characteristics and limitations. While many soft actuators use the unique characteristics of smart materials to achieve certain deformations, certain smart materials have certain limitations and shortcomings. Table 3 shows the characteristics and limitations of a number of common materials.

Table 3. The characteristics and limitations of commonly used soft materials.

Materials	Characteristics	Limitations
Flexible silicone	High expansion coefficient, Long lifetime, Good flexibility and safety.	Long molding time, Complex production method, Expensive.
SMA/SMP	Rich phase transition phenomenon, Excellent shape memory and super elastic property, Good mechanical property, Corrosion resistance.	Complex manufacturing process, High-cost.
MRF/ERF	High boiling point, Low freezing point, Large viscosity change, Good chemical stability, Low-cost.	Poor sedimentation stability.
MRE	Large viscosity change, Good stability, Good sedimentation.	Insufficient liquidity.
DE	Large deformation and fast response, Light weight, High energy density.	High actuation voltage, Low safety.
IPMC	Low driving voltage, Fast response, Low power consumption, Low density and good flexibility.	Back relaxation effects, Low control accuracy.

4. Manufacturing

4.1. Shape Deposition Manufacturing

Shape deposition manufacturing (SDM) is another rapid manufacturing method; after rapid prototyping, it adopts a unique method of adding and removing materials [111–114]. Meanwhile, different methods can be introduced according to the materials used for the parts, making it possible to integrate multiple materials into a single actuator [115]. As shown in Figure 14a, through multi-step deposition and removal of materials, multi-material soft actuators with certain outer shapes and internal structures can be obtained. Moreover, rigid electronic components such as sensors can be embedded in soft materials, which provides soft actuators with a flexible appearance and avoids the confines of rigid materials. Currently, SDM has been widely used in soft actuators manufacturing due to the advantages of low cost, simplicity, and a relatively speedy manufacturing process [116–119]. Similarly, the silicone-molding method and SDM manufacturing method have many common characteristics. The silicone shape can be formed by designing the mold, and the soft actuator can be manufactured by multi-step silicone-molding [120,121].

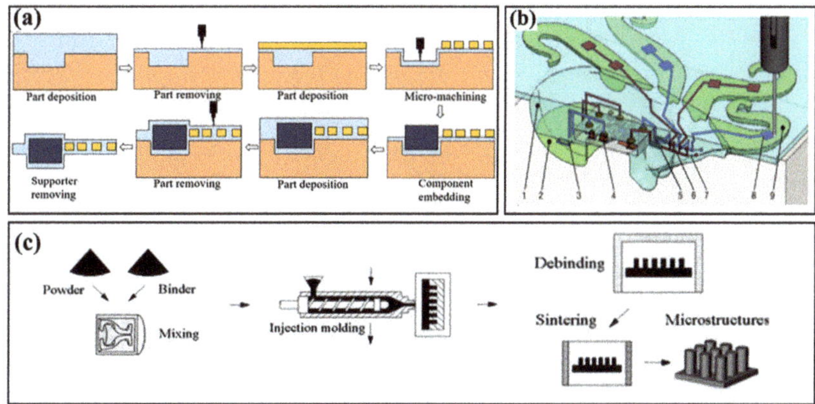

Figure 14. (**a**) Schematic diagram of SDM processes [115]. (**b**) Fully soft robot manufactured by 3D printing [122]. (**c**) Processing steps of micro-powder injection molding [123].

4.2. Three-Dimensional Printing

Three-dimensional (3D) printing is a kind of rapid prototyping technology, sometimes known as additive manufacturing. It is a technology that uses adhesive materials to print objects layer by layer based on digital model files, which is suitable for printing complex structures [124–127]. In the process of printing, the influence of various parameters needs to be considered [128–130]. Early 3D printing technology could only print rigid materials; soft materials were not available in the past, as only thin film parts could be printed due to the viscosity and fluidity of the materials. Wehner et al. made a breakthrough in moving research from semi-soft to fully soft robots [122], as shown in Figure 14b. Using multi-material 3D printing technology, they developed a soft octopus robot able to craw and swim.

4.3. Micro-Powder Injection Molding

Micro-powder injection molding is a new forming technology which combines traditional powder metallurgy technology with modern plastic injection molding technology. It is widely used to produce products with small size and complex shape; therefore, it has been introduced into soft robots as a promising molding method [123]. Soft materials are injected into a mold with a specific shape, then the desired soft parts are obtained through demolding and sintering, as shown in Figure 14c. This method makes it easy to manufacture soft parts with small size [131–134]. However, various defects, such as cracks and bubbles, may appear during the production process. In addition, various parameters which may affect the final molding effect need to be considered, such as the mold temperature, injection pressure, and pressure holding time.

4.4. Soft Lithography

Soft lithography, or soft etching, is a fundamental technology applied to fabrication and microfabrication. It utilizes stamps or molds made from PDMS with microscopic patterns on the surface to achieve microstructure replication by molding and embossing elastomers on a mold [135]. Soft lithography is characterized by replacing the hard mold used in lithography with an elastic mold; thus, it is able to produce more complex three-dimensional structures. In addition, it can change the chemical properties of the material surface as needed, which provieds an opportunity to fabricate multimaterial soft robots [80]. Currently, soft lithography is widely used in optics, biotechnology, microelectronics, sensors, and micro-total analysis systems.

5. Applications

5.1. Wearable Devices

Wearable devices is a general term for a class of products attached to or worn on users as wearable technology, and is commonly used in monitoring and medical fields. Such devices have to be flexible, lightwight, portable, and able to output relatively large torque when used in the medical assistant field. As a result, soft robots are quite suitable for wearable device because of their high compliance, adaptiveness, and safety in human–machine interaction [136,137]. For example, a fluid-driven exoskeleton is designed to be worn on the limb and help disabled people with rehabilitation training, such as standing up independently [138]. Moerover, a flexible exoskeleton with a control system was designed to assist people with walking [139]. However, this flexible exoskeleton cannot provide enough output force and support, making it only suitable for people with a certain level of walking ability. In addition, origami structures have been applied to wearable devices; because of their light weight, actuators made from origami structures can help people who have lost the ability to grip objects with certain weight and size through pneumatic actuation [140]. Furthermore, a low-cost soft neuroprosthetic hand was designed by [141]. In addition to achieving precise movement, the soft neuroprosthetic hand was able to realize particular tactile feedback. Amputees are able to gain real-time feedback from the prosthetic hand, whcih is suitable to wear with only 256 g.

5.2. Exploration and Rescue

Over the past decades, robots have drawn more and more attention in the field of exploration and rescue. Because the working environment is full of complexity and uncertainty, robots must have good anti-damage ability and be able to work normally in an unstructured environment. In addition, they must be miniaturized and lightweight as much as possible in order to meet the requirements of working in a narrow space. Therefore, soft robots have become a research hotspot because of their good environmental adaptability and portability. One example is a novel soft robot able to grow tens of times longer when pressure is applied to its tip, which invests itself with good obstacle avoidance and high load capacity (100 kg) [142]. Cameras and sensors that installed on such robots can achieve real-time transmission of images and signals, which helps them to achieve turning and automatic obstacle avoidance by adjusting the gas flow, which could be useful in earthquake rescue applications. Furthermore, a team from Zhejiang University designed a wireless self-powered soft robot that could be used for deep-sea exploration [143]. Their robot has been tested in the Marianas trench and South China Sea, proving that it can be used in deep sea exploration thanks to its excellent pressure resistance and swimming performance. In addition, soft robots can be used underground. For instance, an underground exploration soft robot has been invented that can burrow in sand at a speed of 480 cm/s [144]. It is used as a carrier for electric wires and irrigation pipes because of its hollow structure and ability to move around and flex under obstacles.

5.3. Industrial Field

Although traditional rigid robots are able to complete automatic operations accurately and efficiently, they may cause potential harm to people because of their rigid structures; in addition, they have shortcomings when dealing with fragile and delicate objects. Therefore, soft robots have attracted much attention in industry because of their good shape adaptability and flexibility [145–152]. One example is a soft gripper based on pneumatic networks that is able to grasp and manipulate objects with complex shapes thanks to its ingenious structural design [106]. Another example is a soft gripper inspired by an octopus that is capable of grasping target objects underwater and in the air through the force provided by suction discs on the gripper or the deformation of the gripper, which has good grasping ability by combining the two approaches [**?**]. A gripper that utilizes particle interference has been designed as well [154]. Coffee beans were inserted into a spherical elastic film to adapt to the shape of objects and variable stiffness grasping was realized by controlling the pressure in the film, with a greater vacuum resulting in higher stiffness. Moreover, a palm with three fingers has been developed that can actively deform to grasp objects with different shapes, which expands the application of soft robots [155]. Additionally, a soft gripper with less control and sensory ability was proposed that is able to manipulate objects without knowing their precise position, shape, or size, including unscrewing caps and sorting snacks [156]. In addition, various hardware facilities can be combined to expand the application of soft robots [157], and grippers assembled on mechanical arms could be used in various industrial tasks.

6. Conclusions and Prospectives

This paper reviews the state-of-the-art research on soft actuators. The principles of various actuations are first summarized. Then, we review different kinds of structural designs used to achieve different motions, such as bending, linear, torsional, spiral, and composite motions. Existing research on manufacturing is discussed, and applications in wearable devices, exploration and rescue, and industrial fields are presented. A variety of actuation methods and motions exist. For different actuation methods, it is necessary to design a suitable preprogrammed structure which controls the deformation of the actuator material in order to achieve a specific motion. The structural design matrix we propose can provide a guide to actuator designs.

Although significant progress has been achieved in soft actuators over the past few decades, many challenges remain in developing soft actuators; these include poor self-healing, low sensory integration, and small region of actuation force. Therefore, soft actuators are expected to improve in the future thanks to their high safety, good controllability, high integration and good repeatability. Meanwhile, the theoretical model of the soft actuator is difficult to model. Non-model behavioral motion control could become a future research direction for researchers; alternatively, other advanced control technologies, such as neural networks, could be used to obtain adaptability to the environment. Furthermore, research of bionic intelligent control algorithms of soft robots should be strengthened. By effectively calculating and controlling the movement, body stiffness, and deformation of soft robots, they could be made to better adapt to changing environments.

Author Contributions: Conceptualization, X.T., P.J., H.W. and Y.Y.; software, H.L.; validation, X.T., H.L., T.M. and J.L.; resources, P.J.; writing—original draft preparation, X.T., H.L. and T.M.; writing—review and editing, X.T.; supervision, P.J.; All authors have read and agreed to the published version of the manuscript.

Funding: The authors gratefully acknowledge financial support from the National Key R&D Program of China (No. 2018YFB1700602), the National Science Foundation of China (No. 52075060, No. 52005269, No. 51705050), the Fundamental Research Funds for the Central Universities (No. 2022CDJKYJH024), and the Natural Science Foundation of Chongqing (No. 2022NSCQ-MSX1629).

Data Availability Statement: Not applicable.

Conflicts of Interest: The authors declare no conflict of interest.

References

1. Brown, E.; Rodenberg, N.; Amend, J.; Mozeika, A.; Steltz, E.; Zakin, M.R.; Lipson, H.; Jaeger, H.M. Universal robotic gripper based on the jamming of granular material. *Proc. Natl. Acad. Sci. USA* **2010**, *107*, 18809–18814. [CrossRef]
2. Yang, Y.; Chen, Y. Innovative Design of Embedded Pressure and Position Sensors for Soft Actuators. *IEEE Robot. Autom. Lett.* **2018**, *3*, 656–663. [CrossRef]
3. Yang, Y.; Chen, Y.; Li, Y.; Wang, Z.; Li, Y. Novel Variable-Stiffness Robotic Fingers with Built-In Position Feedback. *Soft Robot.* **2017**, *4*, 338–352. [CrossRef] [PubMed]
4. Kim, Y.J.; Cheng, S.; Kim, S.; Iagnemma, K. A novel layer jamming mechanism with tunable stiffness capability for minimally invasive surgery. *IEEE Trans. Robot.* **2013**, *29*, 1031–1042. [CrossRef]
5. Dogangil, G.; Davies, B.L.; Baena, F.R. A review of medical robotics for minimally invasive soft tissue surgery. *Proc. Inst. Mech. Eng. Part J. Eng. Med.* **2010**, *224*, 653–679. [CrossRef]
6. Degani, A.; Choset, H.; Wolf, A.; Zenati, M. Highly articulated robotic probe for minimally invasive surgery. In Proceedings of the 2006 IEEE International Conference on Robotics and Automation, ICRA, Orlando, FL, USA, 15–19 May 2006; pp. 4167–4172. [CrossRef]
7. Gao, X.; Sozumert, E.; Zhao, W.W.; Shi, Z.J.; Silberschmidt, V.V. 2-Deformation and fracture behaviors of long-fiber hydrogels. In *The Mechanics of Hydrogels*; Li, H., Silberschmidt, V., Eds.; Elsevier Series in Mechanics of Advanced Materials; Woodhead Publishing: Sawston, UK, 2022; pp. 25–40. [CrossRef]
8. Zhang, C.; Zhu, P.; Lin, Y.; Tang, W.; Jiao, Z.; Yang, H.; Zou, J. Fluid-driven artificial muscles: bio-design, manufacturing, sensing, control, and applications. *Bio Des. Manuf.* **2021**, *4*, 123–145. [CrossRef]
9. Li, Y.; Chen, Y.; Ren, T.; Choi, S.H. Precharged pneumatic soft actuators and their applications to untethered soft robots. *Soft Robot.* **2018**, *5*, 567–575. [CrossRef] [PubMed]
10. Li, Y.; Chen, Y.; Li, Y. Pre-charged pneumatic soft gripper with closed-loop control. *IEEE Robot. Autom. Lett.* **2019**, *4*, 1402–1408. [CrossRef]
11. Cui, Y.; Liu, X.J.; Dong, X.; Zhou, J.; Zhao, H. Enhancing the Universality of a Pneumatic Gripper via Continuously Adjustable Initial Grasp Postures. *IEEE Trans. Robot.* **2021**, *37*, 1604–1618. [CrossRef]
12. Ranzani, T.; Gerboni, G.; Cianchetti, M.; Menciassi, A. A bioinspired soft manipulator for minimally invasive surgery. *Bioinspiration Biomimetics* **2015**, *10*, 035008. [CrossRef]
13. Ren, T.; Li, Y.; Xu, M.; Li, Y.; Xiong, C.; Chen, Y. A Novel Tendon-Driven Soft Actuator with Self-Pumping Property. *Soft Robot.* **2020**, *7*, 130–139. [CrossRef] [PubMed]
14. Shan, W.; Diller, S.; Tutcuoglu, A.; Majidi, C. Rigidity-tuning conductive elastomer. *Smart Mater. Struct.* **2015**, *24*, 065001. [CrossRef]
15. Rich, S.; Jang, S.H.; Park, Y.L.; Majidi, C. Liquid metal-conductive thermoplastic elastomer integration for low-voltage stiffness tuning. *Adv. Mater. Technol.* **2017**, *2*, 1700179. [CrossRef]

16. Nasab, A.M.; Sabzehzar, A.; Tatari, M.; Majidi, C.; Shan, W. A soft gripper with rigidity tunable elastomer strips as ligaments. *Soft Robot.* **2017**, *4*, 411–420. [CrossRef]
17. Cheng, N.G.; Gopinath, A.; Wang, L.; Iagnemma, K.; Hosoi, A.E. Thermally tunable, self-healing composites for soft robotic applications. *Macromol. Mater. Eng.* **2014**, *299*, 1279–1284. [CrossRef]
18. Wang, Q.; Tian, X.; Li, D. Multimodal soft jumping robot with self-decision ability. *Smart Mater. Struct.* **2021**, *30*, 085038. [CrossRef]
19. Yuan, X.; Zhou, X.; Liang, Y.; Wang, L.; Chen, R.; Zhang, M.; Pu, H.; Xuan, S.; Wu, J.; Wen, W. A stable high-performance isotropic electrorheological elastomer towards controllable and reversible circular motion. *Compos. Part B Eng.* **2020**, *193*, 107988. [CrossRef]
20. Yin, L.J.; Zhao, Y.; Zhu, J.; Yang, M.; Zhao, H.; Pei, J.Y.; Zhong, S.L.; Dang, Z.M. Soft, tough, and fast polyacrylate dielectric elastomer for non-magnetic motor. *Nat. Commun.* **2021**, *12*, 4517. [CrossRef]
21. Dang, Z.m.; Wang, L.; Wang, H.y. Novel smart materials: progress in electroactive polymers. *J. Funct. Mater.* **2005**, *36*, 981.
22. Lai, W.; Bastawros, A.F.; Hong, W.; Chung, S.J. Fabrication and analysis of planar dielectric elastomer actuators capable of complex 3-D deformation. In Proceedings of the 2012 IEEE International Conference on Robotics and Automation, Saint Paul, MN, USA, 14–18 May 2012; pp. 4968–4973.
23. Su, Y.; Ye, X.; Guo, S. An autonomous micro robot fish based on IPMC actuator. *Robot* **2010**, *32*, 262–270. [CrossRef]
24. Somlor, S.; Dominguez, G.A.; Schmitz, A.; Kamezaki, M.; Sugano, S. A haptic interface with adjustable stiffness using MR fluid. In Proceedings of the 2015 IEEE International Conference on Advanced Intelligent Mechatronics (AIM), Busan, Korea, 7–11 July 2015; pp. 1132–1137.
25. Guðmundsson, Í. A feasibility study of magnetorheological elastomers for a potential application in prosthetic devices. Ph.D. Thesis, University of Iceland, Reykjavík, Iceland, 2011 .
26. Shembekar, S.; Kamezaki, M.; Zhang, P.; He, Z.; Tsunoda, R.; Otsuki, K.; Sakamoto, H.; Sugano, S. Preliminary Development of a Powerful and Backdrivable Robot Gripper Using Magnetorheological Fluid. In Proceedings of the ISARC—International Symposium on Automation and Robotics in Construction, Kitakyshu, Japan, 27–28 October 2020; Volume 37, pp. 1458–1463.
27. Kashima, S.; Miyasaka, F.; Hirata, K. Novel soft actuator using magnetorheological elastomer. *IEEE Trans. Magn.* **2012**, *48*, 1649–1652. [CrossRef]
28. Hribar, K.C.; Metter, R.B.; Ifkovits, J.L.; Troxler, T.; Burdick, J.A. Light-induced temperature transitions in biodegradable polymer and nanorod composites. *Small* **2009**, *5*, 1830–1834. [CrossRef] [PubMed]
29. Sugiura, S.; Sumaru, K.; Ohi, K.; Hiroki, K.; Takagi, T.; Kanamori, T. Photoresponsive polymer gel microvalves controlled by local light irradiation. *Sens. Actuators Phys.* **2007**, *140*, 176–184. [CrossRef]
30. Rogóż, M.; Zeng, H.; Xuan, C.; Wiersma, D.S.; Wasylczyk, P. Light-driven soft robot mimics caterpillar locomotion in natural scale. *Adv. Opt. Mater.* **2016**, *4*, 1689–1694. [CrossRef]
31. Habibi, Y.; Lucia, L.A.; Rojas, O.J. Cellulose nanocrystals: Chemistry, self-assembly, and applications. *Chem. Rev.* **2010**, *110*, 3479–3500. [CrossRef] [PubMed]
32. Okuzaki, H.; Kuwabara, T.; Kunugi, T. Theoretical study of sorption-induced bending of polypyrrole films. *J. Polym. Sci. Part Polym. Phys.* **1998**, *36*, 2237–2246. [CrossRef]
33. Sidorenko, A.; Krupenkin, T.; Taylor, A.; Fratzl, P.; Aizenberg, J. Reversible switching of hydrogel-actuated nanostructures into complex micropatterns. *Science* **2007**, *315*, 487–490. [CrossRef]
34. Zhang, L.; Liang, H.; Jacob, J.; Naumov, P. Erratum: Photogated humidity-driven motility. *Nat. Commun.* **2015**, *6*, 7862. [CrossRef]
35. Maeda, S.; Hara, Y.; Yoshida, R.; Hashimoto, S. Self-oscillating gel actuator for chemical robotics. *Adv. Robot.* **2008**, *22*, 1329–1342. [CrossRef]
36. Foresti, D.; Nabavi, M.; Klingauf, M.; Ferrari, A.; Poulikakos, D. Acoustophoretic contactless transport and handling of matter in air. *Proc. Natl. Acad. Sci. USA* **2013**, *110*, 12549–12554. [CrossRef]
37. Ding, X.; Lin, S.C.S.; Kiraly, B.; Yue, H.; Li, S.; Chiang, I.K.; Shi, J.; Benkovic, S.J.; Huang, T.J. On-chip manipulation of single microparticles, cells, and organisms using surface acoustic waves. *Proc. Natl. Acad. Sci. USA* **2012**, *109*, 11105–11109. [CrossRef] [PubMed]
38. Laurell, T.; Petersson, F.; Nilsson, A. Chip integrated strategies for acoustic separation and manipulation of cells and particles. *Chem. Soc. Rev.* **2007**, *36*, 492–506. [CrossRef] [PubMed]
39. Zhang, Y.; Yang, D.; Yan, P.; Zhou, P.; Zou, J.; Gu, G. Inchworm Inspired Multimodal Soft Robots With Crawling, Climbing, and Transitioning Locomotion. *IEEE Trans. Robot.* **2021**, *38*, 1806–1819. [CrossRef]
40. Yoshida, S.; Morimoto, Y.; Zheng, L.; Onoe, H.; Takeuchi, S. Multipoint bending and shape retention of a pneumatic bending actuator by a variable stiffness endoskeleton. *Soft Robot.* **2018**, *5*, 718–725. [CrossRef] [PubMed]
41. Yoshida, K.T.; Ren, X.; Blumenschein, L.H.; Okamura, A.M.; Luo, M. AFREEs: Active Fiber Reinforced Elastomeric Enclosures. In Proceedings of the 2020 3rd IEEE International Conference on Soft Robotics (RoboSoft), New Haven, CT, USA, 15 May–15 July 2020; pp. 305–311.
42. Yang, B.; Baines, R.; Shah, D.; Patiballa, S.; Thomas, E.; Venkadesan, M.; Kramer-Bottiglio, R. Reprogrammable soft actuation and shape-shifting via tensile jamming. *Sci. Adv.* **2021**, *7*, eabh2073. [CrossRef]
43. Cui, J.; Huang, T.Y.; Luo, Z.; Testa, P.; Gu, H.; Chen, X.Z.; Nelson, B.J.; Heyderman, L.J. Nanomagnetic encoding of shape-morphing micromachines. *Nature* **2019**, *575*, 164–168. [CrossRef]

44. Ilievski, F.; Mazzeo, A.D.; Shepherd, R.F.; Chen, X.; Whitesides, G.M. Soft robotics for chemists. *Angew. Chem.* **2011**, *123*, 1930–1935. [CrossRef]
45. Lotfiani, A.; Yi, X.; Qinzhi, Z.; Shao, Z.; Wang, L. Design and Experiment of a Fast-Soft Pneumatic Actuator with High Output Force. In Proceedings of the International Conference on Intelligent Robotics and Applications, Newcastle, NSW, Australia, 9–11 August 2018; pp. 442–452.
46. Marchese, A.D.; Katzschmann, R.K.; Rus, D. A recipe for soft fluidic elastomer robots. *Soft Robot.* **2015**, *2*, 7–25. [CrossRef]
47. Gong, X.; Yang, K.; Xie, J.; Wang, Y.; Kulkarni, P.; Hobbs, A.S.; Mazzeo, A.D. Rotary actuators based on pneumatically driven elastomeric structures. *Adv. Mater.* **2016**, *28*, 7533–7538. [CrossRef]
48. Nakai, H.; Kuniyoshi, Y.; Inaba, M.; Inoue, H. Metamorphic robot made of low melting point alloy. In Proceedings of the IEEE/RSJ International Conference on Intelligent Robots and Systems, Lausanne, Switzerland, 30 September–4 October 2002; Volume 2, pp. 2025–2030.
49. Shintake, J.; Schubert, B.; Rosset, S.; Shea, H.; Floreano, D. Variable stiffness actuator for soft robotics using dielectric elastomer and low-melting-point alloy. In Proceedings of the 2015 IEEE/RSJ International Conference on Intelligent Robots and Systems (IROS), Hamburg, Germany, 28 September–2 October 2015; pp. 1097–1102.
50. Yang, Y.; Tse, Y.A.; Zhang, Y.; Kan, Z.; Wang, M.Y. A Low-cost Inchworm-inspired Soft Robot Driven by Supercoiled Polymer Artificial Muscle. In Proceedings of the 2019 2nd IEEE International Conference on Soft Robotics (RoboSoft), Seoul, Korea, 14–18 April 2019; pp. 161–166. [CrossRef]
51. Shaw, J.A.; Kyriakides, S. Thermomechanical aspects of NiTi. *J. Mech. Phys. Solids* **1995**, *43*, 1243–1281. [CrossRef]
52. Kim, B.; Lee, M.G.; Lee, Y.P.; Kim, Y.; Lee, G. An earthworm-like micro robot using shape memory alloy actuator. *Sens. Actuators Phys.* **2006**, *125*, 429–437. [CrossRef]
53. Chang, B.C.; Shaw, J.A.; Iadicola, M.A. Thermodynamics of shape memory alloy wire: modeling, experiments, and application. *Contin. Mech. Thermodyn.* **2006**, *18*, 83–118. [CrossRef]
54. Liu, C.; Qin, H.; Mather, P. Review of progress in shape-memory polymers. *J. Mater. Chem.* **2007**, *17*, 1543–1558. [CrossRef]
55. Firouzeh, A.; Salerno, M.; Paik, J. Stiffness control with shape memory polymer in underactuated robotic origamis. *IEEE Trans. Robot.* **2017**, *33*, 765–777. [CrossRef]
56. Mailen, R.W.; Wagner, C.H.; Bang, R.S.; Zikry, M.; Dickey, M.D.; Genzer, J. Thermo-mechanical transformation of shape memory polymers from initially flat discs to bowls and saddles. *Smart Mater. Struct.* **2019**, *28*, 045011. [CrossRef]
57. Wu, S.; Baker, G.L.; Yin, J.; Zhu, Y. Fast Thermal Actuators for Soft Robotics. *Soft Robot.* **2021**, *5*, 513–527 . [CrossRef]
58. Yang, Y.; Chen, Y. 3D printing of smart materials for robotics with variable stiffness and position feedback. In Proceedings of the 2017 IEEE International Conference on Advanced Intelligent Mechatronics (AIM), Munich, Germany, 3–7 July 2017; pp. 418–423. [CrossRef]
59. Mäkelä, K.K. Characterization and performance of electrorheological fluids based on pine oils. *J. Intell. Mater. Syst. Struct.* **1999**, *10*, 609–614. [CrossRef]
60. Sadeghi, A.; Beccai, L.; Mazzolai, B. Innovative soft robots based on electro-rheological fluids. In Proceedings of the 2012 IEEE/RSJ International Conference on Intelligent Robots and Systems, Vilamoura-Algarve, Portugal, 7–12 October 2012; pp. 4237–4242.
61. Zhu, F.B.; Zhang, C.L.; Qian, J.; Chen, W.Q. Mechanics of dielectric elastomers: materials, structures, and devices. *J. Zhejiang Univ. Sci. A* **2016**, *17*, 1–21.
62. Cao, C.; Chen, L.; Hill, T.L.; Wang, L.; Gao, X. Exploiting Bistability for High-Performance Dielectric Elastomer Resonators. *IEEE/ASME Trans. Mechatron.* **2022**, *15*, 1–12 . [CrossRef]
63. Kellaris, N.; Gopaluni Venkata, V.; Smith, G.M.; Mitchell, S.K.; Keplinger, C. Peano-HASEL actuators: Muscle-mimetic, electrohydraulic transducers that linearly contract on activation. *Sci. Robot.* **2018**, *3*, eaar3276. [CrossRef]
64. Bahramzadeh, Y.; Shahinpoor, M. A review of ionic polymeric soft actuators and sensors. *Soft Robot.* **2014**, *1*, 38–52. [CrossRef]
65. Fujita, T.; Shimada, K. Characteristics and applications of magnetorheological fluids. *J. Magn. Soc. Jpn.* **2003**, *27*, 91–100.
66. Wang, X.; Gordaninejad, F.; Calgar, M.; Liu, Y.; Sutrisno, J.; Fuchs, A. Sensing behavior of magnetorheological elastomers. *J. Mech. Des.* **2009**, *131*, 091004. [CrossRef]
67. Hiraki, T.; Nakahara, K.; Narumi, K.; Niiyama, R.; Kida, N.; Takamura, N.; Okamoto, H.; Kawahara, Y. Laser pouch motors: Selective and wireless activation of soft actuators by laser-powered liquid-to-gas phase change. *IEEE Robot. Autom. Lett.* **2020**, *5*, 4180–4187. [CrossRef]
68. Lendlein, A.; Jiang, H.; Jünger, O.; Langer, R. Light-induced shape-memory polymers. *Nature* **2005**, *434*, 879–882. [CrossRef]
69. Ma, M.; Guo, L.; Anderson, D.G.; Langer, R. Bio-inspired polymer composite actuator and generator driven by water gradients. *Science* **2013**, *339*, 186–189. [CrossRef]
70. Zhang, K.; Geissler, A.; Standhardt, M.; Mehlhase, S.; Gallei, M.; Chen, L.; Thiele, C.M. Moisture-responsive films of cellulose stearoyl esters showing reversible shape transitions. *Sci. Rep.* **2015**, *5*, 11011. [CrossRef]
71. Bartlett, N.W.; Tolley, M.T.; Overvelde, J.T.; Weaver, J.C.; Mosadegh, B.; Bertoldi, K.; Whitesides, G.M.; Wood, R.J. A 3D-printed, functionally graded soft robot powered by combustion. *Science* **2015**, *349*, 161–165. [CrossRef]
72. Onal, C.D.; Chen, X.; Whitesides, G.M.; Rus, D. Soft mobile robots with on-board chemical pressure generation. In *Robotics Research*; Springer: Cham, Switzerland, 2017; pp. 525–540.
73. Whymark, R. Acoustic field positioning for containerless processing. *Ultrasonics* **1975**, *13*, 251–261. [CrossRef]

74. Yang, Y.; Li, Y.; Chen, Y. Principles and methods for stiffness modulation in soft robot design and development. *Bio Des. Manuf.* **2018**, *1*, 14–25. [CrossRef]
75. Wang, Z.; Polygerinos, P.; Overvelde, J.T.; Galloway, K.C.; Bertoldi, K.; Walsh, C.J. Interaction forces of soft fiber reinforced bending actuators. *IEEE/ASME Trans. Mechatronics* **2016**, *22*, 717–727. [CrossRef]
76. Galloway, K.C.; Polygerinos, P.; Walsh, C.J.; Wood, R.J. Mechanically programmable bend radius for fiber-reinforced soft actuators. In Proceedings of the 2013 16th International Conference on Advanced Robotics (ICAR), Montevideo, Uruguay, 25–29 November 2013; pp. 1–6. [CrossRef]
77. Ye, Y.; Cheng, P.; Yan, B.; Lu, Y.; Wu, C. Design of a Novel Soft Pneumatic Gripper with Variable Gripping Size and Mode. *J. Intell. Robot. Syst.* **2022**, *106*, 5. [CrossRef]
78. Deimel, R.; Brock, O. A novel type of compliant and underactuated robotic hand for dexterous grasping. *Int. J. Robot. Res.* **2016**, *35*, 161–185. [CrossRef]
79. Mosadegh, B.; Polygerinos, P.; Keplinger, C.; Wennstedt, S.; Shepherd, R.F.; Gupta, U.; Shim, J.; Bertoldi, K.; Walsh, C.J.; Whitesides, G.M. Pneumatic Networks for Soft Robotics that Actuate Rapidly. *Adv. Funct. Mater.* **2014**, *24*, 2163–2170. [CrossRef]
80. Shepherd, R.F.; Ilievski, F.; Choi, W.; Morin, S.A.; Stokes, A.A.; Mazzeo, A.D.; Chen, X.; Wang, M.; Whitesides, G.M. Multigait soft robot. *Proc. Natl. Acad. Sci. USA* **2011**, *108*, 20400–20403. [CrossRef]
81. Kim, W.; Seo, B.; Yu, S.Y.; Cho, K.J. Deployable Soft Pneumatic Networks (D-PneuNets) Actuator with Dual-Morphing Origami Chambers for High-Compactness. *IEEE Robot. Autom. Lett.* **2022**, *7*, 1262–1269. [CrossRef]
82. Kitano, S.; Komatsuzaki, T.; Suzuki, I.; Nogawa, M.; Naito, H.; Tanaka, S. Development of a rigidity tunable flexible joint using magneto-rheological compounds—Toward a multijoint manipulator for laparoscopic surgery. *Front. Robot. AI* **2020**, *7*, 59. [CrossRef]
83. Yang, Y.; Chen, Y. Novel design and 3D printing of variable stiffness robotic fingers based on shape memory polymer. In Proceedings of the 2016 6th IEEE International Conference on Biomedical Robotics and Biomechatronics (BioRob), Singapore, 26–29 June 2016; pp. 195–200.
84. Yuk, H.; Kim, D.; Lee, H.; Jo, S.; Shin, J.H. Shape memory alloy-based small crawling robots inspired by *C. elegans*. *Bioinspiration Biomimetics* **2011**, *6*, 046002. [CrossRef]
85. Li, T.; Li, G.; Liang, Y.; Cheng, T.; Dai, J.; Yang, X.; Liu, B.; Zeng, Z.; Huang, Z.; Luo, Y.; et al. Fast-moving soft electronic fish. *Sci. Adv.* **2017**, *3*, e1602045. [CrossRef]
86. Yang, Y.; Chen, Y.; Li, Y.; Chen, M.Z.; Wei, Y. Bioinspired Robotic Fingers Based on Pneumatic Actuator and 3D Printing of Smart Material. *Soft Robot.* **2017**, *4*, 147–162. [CrossRef]
87. Zhang, X.; Yu, Z.; Wang, C.; Zarrouk, D.; Seo, J.W.; Cheng, J.; Buchan, A.; Takei, K.; Zhao, Y.; Ager, J.; et al. Photoactuators and motors based on carbon nanotubes with selective chirality distributions. *Nat. Commun.* **2014**, *5*, 2983. [CrossRef] [PubMed]
88. Gavrilović, M.M.; Marić, M.R. Positional servo-mechanism activated by artificial muscles. *Med. Biol. Eng.* **1969**, *7*, 77–82. [CrossRef] [PubMed]
89. Yamaguchi, A.; Takemura, K.; Yokota, S.; Edamura, K. An in-pipe mobile robot using electro-conjugate fluid. *J. Adv. Mech. Des. Syst. Manuf.* **2011**, *5*, 214–226. [CrossRef]
90. Menciassi, A.; Gorini, S.; Pernorio, G.; Dario, P. A SMA actuated artificial earthworm. In Proceedings of the IEEE International Conference on Robotics and Automation, ICRA'04, New Orleans, LA, USA, 26 April–1 May 2004; Volume 4, pp. 3282–3287.
91. Ni, D.; Heisser, R.; Davaji, B.; Ivy, L.; Shepherd, R.; Lal, A. Polymer interdigitated pillar electrostatic (PIPE) actuators. *Microsys. Nanoeng.* **2022**, *8*, 1–8. [CrossRef]
92. Hashem, R.; Stommel, M.; Cheng, L.; Xu, W. Design and characterization of a bellows-driven soft pneumatic actuator. *IEEE/ASME Trans. Mechatron.* **2020**, *26*, 2327–2338. [CrossRef]
93. Li, S.; Vogt, D.M.; Rus, D.; Wood, R.J. Fluid-driven origami-inspired artificial muscles. *Proc. Natl. Acad. Sci. USA* **2017**, *114*, 13132–13137. [CrossRef]
94. Zhang, Z.; Chen, G.; Wu, H.; Kong, L.; Wang, H. A pneumatic/cable-driven hybrid linear actuator with combined structure of origami chambers and deployable mechanism. *IEEE Robot. Autom. Lett.* **2020**, *5*, 3564–3571. [CrossRef]
95. Aziz, S.; Villacorta, B.; Naficy, S.; Salahuddin, B.; Gao, S.; Baigh, T.A.; Sangian, D.; Zhu, Z. A microwave powered polymeric artificial muscle. *Appl. Mater. Today* **2021**, *23*, 101021. [CrossRef]
96. Gorissen, B.; Chishiro, T.; Shimomura, S.; Reynaerts, D.; De Volder, M.; Konishi, S. Flexible pneumatic twisting actuators and their application to tilting micromirrors. *Sens. Actuators A Phys.* **2014**, *216*, 426–431. [CrossRef]
97. Xiao, W.; Du, X.; Chen, W.; Yang, G.; Hu, D.; Han, X. Cooperative collapse of helical structure enables the actuation of twisting pneumatic artificial muscle. *Int. J. Mech. Sci.* **2021**, *201*, 106483. [CrossRef]
98. Jiao, Z.; Zhang, C.; Ruan, J.; Tang, W.; Lin, Y.; Zhu, P.; Wang, J.; Wang, W.; Yang, H.; Zou, J. Re-foldable origami-inspired bidirectional twisting of artificial muscles reproduces biological motion. *Cell Rep. Phys. Sci.* **2021**, *2*, 100407. [CrossRef]
99. Shim, J.E.; Quan, Y.J.; Wang, W.; Rodrigue, H.; Song, S.H.; Ahn, S.H. A smart soft actuator using a single shape memory alloy for twisting actuation. *Smart Mater. Struct.* **2015**, *24*, 125033. [CrossRef]
100. Ahn, S.H.; Lee, K.T.; Kim, H.J.; Wu, R.; Kim, J.S.; Song, S.H. Smart soft composite: An integrated 3D soft morphing structure using bend-twist coupling of anisotropic materials. *Int. J. Precis. Eng. Manuf.* **2012**, *13*, 631–634. [CrossRef]
101. Lee, J.Y.; Kim, W.B.; Choi, W.Y.; Cho, K.J. Soft robotic blocks: Introducing SoBL, a fast-build modularized design block. *IEEE Robot. Autom. Mag.* **2016**, *23*, 30–41. [CrossRef]

102. Schaffner, M.; Faber, J.A.; Pianegonda, L.; Rühs, P.A.; Coulter, F.; Studart, A.R. 3D printing of robotic soft actuators with programmable bioinspired architectures. *Nat. Commun.* **2018**, *9*, 878. [CrossRef] [PubMed]
103. Yan, J.; Zhang, X.; Xu, B.; Zhao, J. A New Spiral-Type Inflatable Pure Torsional Soft Actuator. *Soft Robot.* **2018**, *5*, 527–540. [CrossRef]
104. Feng, Y.; Xu, J.; Zeng, S.; Gao, Y.; Tan, J. Controlled helical deformation of programmable bilayer structures: design and fabrication. *Smart Mater. Struct.* **2020**, *29*, 085042. [CrossRef]
105. Hoang, T.T.; Phan, P.T.; Thai, M.T.; Lovell, N.H.; Do, T.N. Bio-Inspired Conformable and Helical Soft Fabric Gripper with Variable Stiffness and Touch Sensing. *Adv. Mater. Technol.* **2020**, *5*, 2000724. [CrossRef]
106. Wang, T.; Ge, L.; Gu, G. Programmable design of soft pneu-net actuators with oblique chambers can generate coupled bending and twisting motions. *Sensors Actuators A Phys.* **2018**, *271*, 131–138. [CrossRef]
107. Amase, H.; Nishioka, Y.; Yasuda, T. Mechanism and basic characteristics of a helical inflatable gripper. In Proceedings of the 2015 IEEE International Conference on Mechatronics and Automation (ICMA), Beijing, China, 2–5 August 2015; pp. 2559–2564.
108. Guan, Q.; Sun, J.; Liu, Y.; Wereley, N.M.; Leng, J. Novel Bending and Helical Extensile/Contractile Pneumatic Artificial Muscles Inspired by Elephant Trunk. *Soft Robot.* **2020**, *7*, 597–614. [CrossRef]
109. Kim, S.Y.; Baines, R.; Booth, J.; Vasios, N.; Bertoldi, K.; Kramer-Bottiglio, R. Reconfigurable soft body trajectories using unidirectionally stretchable composite laminae. *Nat. Commun.* **2019**, *10*, 3464. [CrossRef] [PubMed]
110. Jiang, P.; Luo, J.; Li, J.; Chen, M.Z.Q.; Chen, Y.; Yang, Y.; Chen, R. A Novel Scaffold-Reinforced Actuator With Tunable Attitude Ability for Grasping. *IEEE Trans. Robot.* **2022**, 12, 1–14 . [CrossRef]
111. Li, X.; Golnas, A.; Prinz, F.B. Shape deposition manufacturing of smart metallic structures with embedded sensors. In *Smart Structures and Materials 2000: Sensory Phenomena and Measurement Instrumentation for Smart Structures and Materials*; International Society for Optics and Photonics: Bellingham, WA, USA, 2000; Volume 3986, pp. 160–171.
112. Merz, R.; Prinz, F.; Ramaswami, K.; Terk, M.; Weiss, L. Shape deposition manufacturing. In Proceedings of the 1994 International Solid Freeform Fabrication Symposium, Austin, TX, USA, 8–10 August 1994.
113. Parenti, P.; Cataldo, S.; Annoni, M. Shape deposition manufacturing of 316L parts via feedstock extrusion and green-state milling. *Manuf. Lett.* **2018**, *18*, 6–11. [CrossRef]
114. Park, Y.L.; Chau, K.; Black, R.J.; Cutkosky, M.R. Force sensing robot fingers using embedded fiber Bragg grating sensors and shape deposition manufacturing. In Proceedings of the 2007 IEEE International Conference on Robotics and Automation, Rome, Italy, 10–14 April 2007; pp. 1510–1516.
115. Cho, K.J.; Koh, J.S.; Kim, S.; Chu, W.S.; Hong, Y.; Ahn, S.H. Review of manufacturing processes for soft biomimetic robots. *Int. J. Precis. Eng. Manuf.* **2009**, *10*, 171–181. [CrossRef]
116. Cham, J.G.; Bailey, S.A.; Clark, J.E.; Full, R.J.; Cutkosky, M.R. Fast and robust: Hexapedal robots via shape deposition manufacturing. *Int. J. Robot. Res.* **2002**, *21*, 869–882. [CrossRef]
117. Suresh, S.A.; Christensen, D.L.; Hawkes, E.W.; Cutkosky, M. Surface and shape deposition manufacturing for the fabrication of a curved surface gripper. *J. Mech. Robot.* **2015**, *7*, 021005. [CrossRef]
118. Gafford, J.; Ding, Y.; Harris, A.; McKenna, T.; Polygerinos, P.; Holland, D.; Walsh, C.; Moser, A. Shape deposition manufacturing of a soft, atraumatic, deployable surgical grasper. *J. Med. Dev.* **2014**, *8*, 021006. [CrossRef]
119. Dollar, A.M.; Howe, R.D. The highly adaptive SDM hand: Design and performance evaluation. *Int. J. Robot. Res.* **2010**, *29*, 585–597. [CrossRef]
120. Dekker, C. Stereolithography tooling for silicone molding. *Adv. Mater. Process.* **2003**, *161*, 59–61.
121. Na, H.K.; Ahn, J.Y.; Lee, G.H.; Lee, J.H.; Kim, D.H.; Jung, K.W.; Choi, K.D.; Song, H.J.; Jung, H.Y. The efficacy of a novel percutaneous endoscopic gastrostomy simulator using three-dimensional printing technologies. *J. Gastroenterol. Hepatol.* **2019**, *34*, 561–566. [CrossRef]
122. Wehner, M.; Truby, R.L.; Fitzgerald, D.J.; Mosadegh, B.; Whitesides, G.M.; Lewis, J.A.; Wood, R.J. An integrated design and fabrication strategy for entirely soft, autonomous robots. *Nature* **2016**, *536*, 451–455. [CrossRef] [PubMed]
123. Fu, G.; Loh, N.; Tor, S.; Murakoshi, Y.; Maeda, R. Replication of metal microstructures by micro powder injection molding. *Mater. Des.* **2004**, *25*, 729–733. [CrossRef]
124. Truby, R.L.; Wehner, M.; Grosskopf, A.K.; Vogt, D.M.; Uzel, S.G.; Wood, R.J.; Lewis, J.A. Soft somatosensitive actuators via embedded 3D printing. *Adv. Mater.* **2018**, *30*, 1706383. [CrossRef] [PubMed]
125. Truby, R.L. Embedded Three-Dimensional Printing of Autonomous and Somatosensitive Soft Robots. Ph.D. Thesis, Harvard University, Cambridge, MA, USA, 2018.
126. Zhu, M.; Xie, M.; Lu, X.; Okada, S.; Kawamura, S. A soft robotic finger with self-powered triboelectric curvature sensor based on multi-material 3D printing. *Nano Energy* **2020**, *73*, 104772. [CrossRef]
127. Xie, M.; Zhu, M.; Yang, Z.; Okada, S.; Kawamura, S. Flexible self-powered multifunctional sensor for stiffness-tunable soft robotic gripper by multimaterial 3D printing. *Nano Energy* **2021**, *79*, 105438. [CrossRef]
128. Truby, R.L.; Lewis, J.A. Printing soft matter in three dimensions. *Nature* **2016**, *540*, 371–378. [CrossRef] [PubMed]
129. Wallin, T.; Pikul, J.; Shepherd, R. 3D printing of soft robotic systems. *Nat. Rev. Mater.* **2018**, *3*, 84–100. [CrossRef]
130. Gul, J.Z.; Sajid, M.; Rehman, M.M.; Siddiqui, G.U.; Shah, I.; Kim, K.H.; Lee, J.W.; Choi, K.H. 3D printing for soft robotics–A review. *Sci. Technol. Adv. Mater.* **2018**, *19*, 243–262. [CrossRef]

131. Attia, U.M.; Alcock, J.R. A review of micro-powder injection moulding as a microfabrication technique. *J. Micromechanics Microeng.* **2011**, *21*, 043001. [CrossRef]
132. Loh, N.; Tor, S.; Tay, B.; Murakoshi, Y.; Maeda, R. Fabrication of micro gear by micro powder injection molding. *Microsyst. Technol.* **2008**, *14*, 43–50. [CrossRef]
133. You, W.K.; Choi, J.P.; Yoon, S.M.; Lee, J.S. Low temperature powder injection molding of iron micro-nano powder mixture. *Powder Technol.* **2012**, *228*, 199–205. [CrossRef]
134. Checot-Moinard, D.; Rigollet, C.; Lourdin, P. Powder injection moulding PIM of feedstock based on hydrosoluble binder and submicronic powder to manufacture parts having micro-details. *Powder Technol.* **2011**, *208*, 472–479. [CrossRef]
135. Onal, C.D.; Rus, D. A modular approach to soft robots. In Proceedings of the 2012 4th IEEE RAS & EMBS International Conference on Biomedical Robotics and Biomechatronics (BioRob), Rome, Italy, 24–27 June 2012; pp. 1038–1045.
136. Majidi, C. Soft robotics: a perspective—current trends and prospects for the future. *Soft Robot.* **2014**, *1*, 5–11. [CrossRef]
137. Banerjee, H.; Tse, Z.T.H.; Ren, H. Soft robotics with compliance and adaptation for biomedical applications and forthcoming challenges. *Int. J. Robot. Autom.* **2018**, *33*, 69–80. [CrossRef]
138. Caldwell, D.G.; Tsagarakis, N.G.; Kousidou, S.; Costa, N.; Sarakoglou, I. "Soft" exoskeletons for upper and lower body rehabilitation— Design, control and testing. *Int. J. Humanoid Robot.* **2007**, *04*, 549–573. [CrossRef]
139. Asbeck, A.T.; De Rossi, S.M.; Holt, K.G.; Walsh, C.J. A biologically inspired soft exosuit for walking assistance. *Int. J. Robot. Res.* **2015**, *34*, 744–762. [CrossRef]
140. Liu, S.; Zhu, Y.; Zhang, Z.; Fang, Z.; Tan, J.; Peng, J.; Song, C.; Asada, H.; Wang, Z. Otariidae-inspired Soft-robotic Supernumerary Flippers by Fabric Kirigami and Origami. *IEEE/ASME Trans. Mechatron.* **2020**, *26*, 2747–2757. [CrossRef]
141. Gu, G.; Zhang, N.; Xu, H.; Lin, S.; Yu, Y.; Chai, G.; Ge, L.; Yang, H.; Shao, Q.; Sheng, X.; et al. A soft neuroprosthetic hand providing simultaneous myoelectric control and tactile feedback. *Nat. Biomed. Eng.* **2021**, 1–10. [CrossRef] [PubMed]
142. Greer, J.D.; Blumenschein, L.H.; Alterovitz, R.; Hawkes, E.W.; Okamura, A.M. Robust navigation of a soft growing robot by exploiting contact with the environment. *Int. J. Robot. Res.* **2020**, *39*, 1724–1738. [CrossRef]
143. Li, G.; Chen, X.; Zhou, F.; Liang, Y.; Xiao, Y.; Cao, X.; Zhang, Z.; Zhang, M.; Wu, B.; Yin, S.; et al. Self-powered soft robot in the Mariana Trench. *Nature* **2021**, *591*, 66–71. [CrossRef]
144. Naclerio, N.D.; Karsai, A.; Murray-Cooper, M.; Ozkan-Aydin, Y.; Aydin, E.; Goldman, D.I.; Hawkes, E.W. Controlling subterranean forces enables a fast, steerable, burrowing soft robot. *Sci. Robot.* **2021**, *6*, eabe2922. [CrossRef] [PubMed]
145. Rus, D.; Tolley, M.T. Design, fabrication and control of soft robots. *Nature* **2015**, *521*, 467–475. [CrossRef] [PubMed]
146. Nishida, T.; Okatani, Y.; Tadakuma, K. Development of universal robot gripper using MR α fluid. *Int. J. Humanoid Robot.* **2016**, *13*, 1650017. [CrossRef]
147. Koivikko, A.; Drotlef, D.M.; Sitti, M.; Sariola, V. Magnetically switchable soft suction grippers. *Extrem. Mech. Lett.* **2021**, *44*, 101263. [CrossRef]
148. Qi, Q.; Xiang, C.; Ho, V.A.; Rossiter, J. A Sea-Anemone-Inspired, Multifunctional, Bistable Gripper. *Soft Robot.* 2021, *Online ahead of print*.
149. Li, Y.; Chen, Y.; Yang, Y.; Wei, Y. Passive Particle Jamming and Its Stiffening of Soft Robotic Grippers. *IEEE Trans. Robot.* **2017**, *33*, 446–455. [CrossRef]
150. Yang, Y.; Kan, Z.; Zhang, Y.; Tse, Y.A.; Wang, M.Y. A Novel Variable Stiffness Actuator Based on Pneumatic Actuation and Supercoiled Polymer Artificial Muscles. In Proceedings of the 2019 International Conference on Robotics and Automation (ICRA), Montreal, QC, Canada, 20–24 May 2019; pp. 3983–3989. [CrossRef]
151. Li, Y.; Tao, R.; Li, Y.; Yang, Y.; Zhou, J.; Chen, Y. A Highly Adaptive Robotic Gripper Palm with Tactile Sensing. In Proceedings of the 2022 IEEE International Conference on Real-time Computing and Robotics (RCAR), Guiyang, China, 17–22 July 2022; pp. 130–135. [CrossRef]
152. Zhou, Y.; Headings, L.M.; Dapino, M.J. Modeling of Fluidic Prestressed Composite Actuators With Application to Soft Robotic Grippers. *IEEE Trans. Robot.* **2022**, *38*, 2166–2178. [CrossRef]
153. oi: 10.1002/advs.202104382 Wu, M.; Zheng, X.; Liu, R.; Hou, N.; Afridi, W.H.; Afridi, R.H.; Guo, X.; Wu, J.; Wang, C.; Xie, G. Glowing Sucker Octopus (Stauroteuthis syrtensis)-Inspired Soft Robotic Gripper for Underwater Self-Adaptive Grasping and Sensing. *Adv. Sci.* **2022**, *9*, 2104382. [CrossRef]
154. Wang, Y.; Yang, Z.; Zhou, H.; Zhao, C.; Barimah, B.; Li, B.; Xiang, C.; Li, L.; Gou, X.; Luo, M. Inflatable Particle-Jammed Robotic Gripper Based on Integration of Positive Pressure and Partial Filling. *Soft Robot.* **2021**, *9*, 309–323. [CrossRef]
155. Subramaniam, V.; Jain, S.; Agarwal, J.; Valdivia y Alvarado, P. Design and characterization of a hybrid soft gripper with active palm pose control. *Int. J. Robot. Res.* **2020**, *39*, 1668–1685. [CrossRef]
156. Abondance, S.; Teeple, C.B.; Wood, R.J. A dexterous soft robotic hand for delicate in-hand manipulation. *IEEE Robot. Autom. Lett.* **2020**, *5*, 5502–5509. [CrossRef]
157. Martinez, R.V.; Branch, J.L.; Fish, C.R.; Jin, L.; Shepherd, R.F.; Nunes, R.M.; Suo, Z.; Whitesides, G.M. Robotic tentacles with three-dimensional mobility based on flexible elastomers. *Adv. Mater.* **2013**, *25*, 205–212. [CrossRef] [PubMed]

Article

Soft Coiled Pneumatic Actuator with Integrated Length-Sensing Function for Feedback Control

Jacob R. Greenwood † and Wyatt Felt *,†

VPI Technology, Draper, UT 84020, USA
* Correspondence: wyattf@vpitech.com
† These authors contributed equally to this work.

Abstract: SPIRA Coil actuators are formed from thin sheets of PET plastic laminated into a coil shape that unfurls like a "party horn" when inflated, while many soft actuators require large pressures to create only modest strains, SPIRA Coils can easily be designed and fabricated to extend over dramatic distances with relatively low working pressures. Internal metalized PET strips separate in the extended portion of the actuator, creating an electrical circuit with a resistance that corresponds to the actuator length. This paper presents and experimentally validates easy-to-use design models for the actuators' self-retracting spring stiffness, its pneumatic extension force, and its internal length-sensing electrical resistance. Testing of the self-sensing capabilities demonstrates that the embedded sensor can be used to determine the actuator length with virtually no hysteresis. Feedback control with the resistance-based sensing resulted in length-control errors within 5% of the extended actuator length (i.e., 3 cm of 60 cm).

Keywords: soft robotics; pneumatic actuators; resistance sensing

Citation: Greenwood, J.R.; Felt, W. Soft Coiled Pneumatic Actuator with Integrated Length-Sensing Function for Feedback Control. *Actuators* **2023**, *12*, 455. https://doi.org/10.3390/act12120455

Academic Editor: Hamed Rahimi Nohooji

Received: 28 October 2023
Revised: 27 November 2023
Accepted: 4 December 2023
Published: 8 December 2023

1. Introduction

The development and application of a wide-variety of soft pneumatic actuation techniques has led to extensive innovation in mechanisms and robotics [1–3]. Typically made from elastomers, fabrics, or thin plastics, these soft actuators are driven by the expansion of an internally contained volume of air. The actuators tend to behave in a "soft" or compliant way because the materials are often flexible and the air is compressible. Examples of common soft pneumatic actuators include axially contracting fiber-reinforced actuators, bellows-like axially extending actuators, and bellows-like bending actuators (which reinforce one side of the actuator to convert the extension into bending).

While these actuators have shown great versatility in their applications, the distance over which these actuators can traverse is fundamentally limited to some fraction or multiple of their initial length. Contracting fiber-reinforced actuators and "peano" actuators, for example, can typically only reduce their length by less than 50% [4,5]. Bellows-like actuators, on the other hand, can extend in an accordion-like manner to several multiples of their contracted length when pressurized. Correspondingly, when subjected to vacuum pressure, bellows-like structures can act as contractile actuators with dramatic contraction ratios [6–8]. The ratio of the extended length to the contracted length is limited, however, by the thickness of the many folds.

In many ways, the extended length of soft pneumatic actuators is limited by the ability to efficiently store the material needed to enclose the air volume as it expands. Some actuators achieve their motion by stretching elastomeric bladders. However, these often require substantial amounts of pressure to stretch the elastomers. Furthermore, there are limits to the strain these materials can undergo. Larger strains can often be accomplished by using membrane-reinforced actuators that use bending or unfolding pleated structures [4,5,7]. Recently, there has been exploration of pneumatically actuated structures that extend by inverting [9,10] or un-spooling thin tubes [11]. These long thin tubes allow the actuators to extend over dramatically long distances.

This paper introduces a novel, self-sensing coiled actuator called a "SPIRA Coil", which stands for Soft Self-Sensing Pneumatic Integrated Retractable Actuator Coil. A SPIRA Coil is formed from thin sheets of plastic that are laminated into a coil shape. When air is added to the interior volume of the actuator, it extends like a party horn (Figure 1b). When air is removed, it retracts towards its coiled shape. Because the materials are so thin when deflated, the actuator can extend and retract over dramatic distances compared to its initial size. In this work, actuators with a coiled outer diameter of approximately 2 cm were able to extend 60 cm.

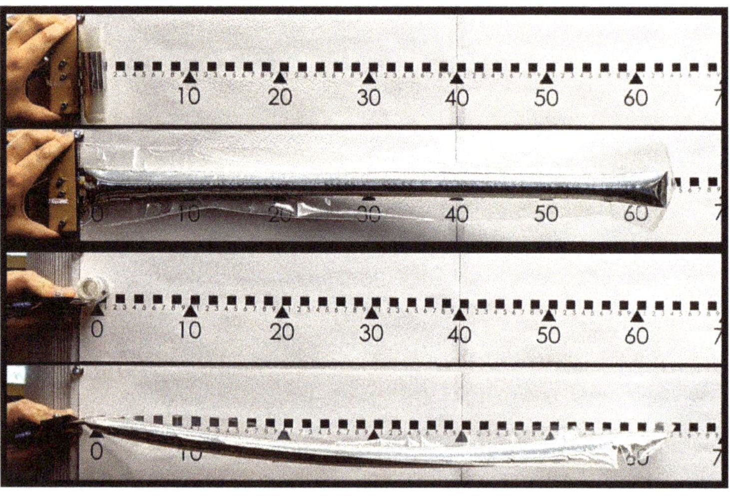

(a)

(b)

Figure 1. The SPIRA Coil actuator presented in this work, (**a**) shown both retracted and extended, is similar in mechanism to a "party horn" (**b**) and can extend to dramatically long lengths from a tightly wound coil. Unlike a party horn, SPIRA Coils can sense their own extension through the resistance of an internal circuit, allowing the length of the actuator to be measured and controlled for robotic applications. The ruler shown in (**a**) is in cm.

Like other soft actuators, the party-horn-like actuator presented in this work could not easily be combined with traditional mechanical sensors without compromising its unique features [12]. Accordingly, SPIRA Coils are designed with sensing integrated into their structure. The interior volume is formed from two layers of PET with a thin metalized layer. The air in the expanded section of the actuator separates these two layers. At the point where the coil begins, the two layers are pressed together, forming an electrical circuit between the air-inlet end of the actuator and the beginning of the coil (Figure 2). As the actuator extends, the length of the circuit increases, making the resistance proportional to the actuator length.

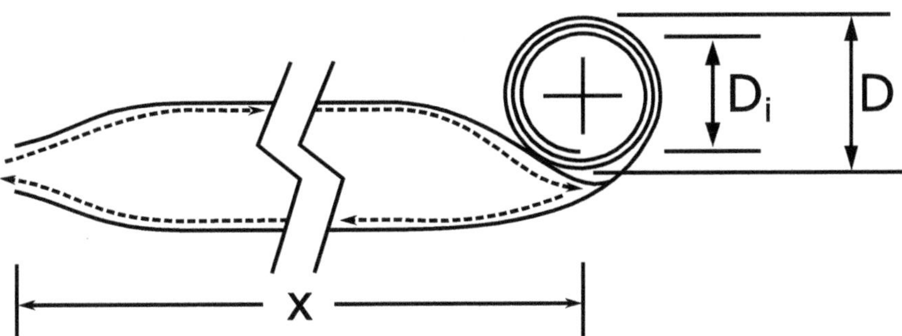

Figure 2. A cross-section of a partially deployed actuator. The conductive path for resistance sensing is shown as a dashed line. Shown also is the corresponding length of the actuator x, the diameter D, and inner diameter D_i.

This paper introduces the SPIRA Coil actuator, describes how it can be readily fabricated with inexpensive materials, and presents and validates easy-to-use design models for the actuator's self-retracting spring stiffness, its pneumatic extension force, and its internal length-sensing electrical resistance. This paper also presents the use of the resistance-based length sensing in the feedback control of the actuators.

2. Background

2.1. Party Horns and Coiled Inflatable Booms

The coiled spiral pneumatic mechanism presented in this work is very similar to that of the "party horn" (Figure 1b). Though the name in English is somewhat ambiguous, other languages have more unique monikers (German—*rollpfeife, luftrüssel*; Spanish—*espantasuegras*; Italian—*lingua di Menelik*; Japanese—*fukimodoshi*) some of which date back more than a century to the popularity of the horn in carnival celebrations. Despite the ubiquity of the mechanism in popular culture, with little exception [13], it has been largely ignored by the robotics community.

In contrast, non-retracting coiled inflatable booms have been the subject of extensive research. These are pneumatic tubes that have been flattened, wrapped around a hub, and designed to unroll as they are inflated [14]. A coiled boom may be advantageous due to its simplicity [15] and the potential for high packing efficiency [16]. Typically inflatable booms, both coiled and uncoiled, use metal–polymer laminates as their membrane [14,15,17–19].

There are, however, known issues with coiled inflatable booms. The thickness of the material, for example, often causes issues (ex. wrinkling, local buckling) during the coiling process [14,17,19,20], while alternate wrapping methods have been suggested that accommodate thickness [21–23]; these methods do not include a sealed inner volume and thus are not suitable for pneumatic actuator use.

The SPIRA Coil fabrication method introduced in this paper solves many of these issues. Rather than rolling up a flat tube, which could introduce wrinkles and local buckling, the layers of the SPIRA Coil prototypes are adhered only after they have been coiled, allowing the layers to shift relative to one another prior to adhesion. The compromise of this method is the slight curvature introduced into the inflated actuator.

Another issue of non-retracting inflatable booms is the potential that they can uncoil in an unpredictable manner. The dynamics and geometry of inflatable booms during deployment have been studied and simulated by [24–28]. Results have found that the deploying motion can be unstable, unpredictable, chaotic, and even catastrophic [15,20,27], particularly in an orbital environment. Researchers have tried to address this by using hook-and-loop strips [29], adhesive [26], and flexible plates [30] as methods to slow down deployment.

In contrast, just like a party horn, the self-retracting SPIRA coil extends and retracts in a predictable way. This is partly encouraged by changes in the relaxed coil diameter along the length of the actuator—an unavoidable result of the fabrication process. This changing diameter, combined with the restricted airflow to the coiled section, creates conditions that encourage the actuator to unfurl sequentially and predictably.

2.2. Plastic Annealing of PET

Polyethylene terephthalate (PET) is a common material used in deployable booms and other origami applications [14,31,32]. It has been demonstrated that heating a sheet of PET can relieve internal stresses without causing softening of the material [33]. This plastic annealing effectively resets the natural state of the PET so that, when the sheet is cooled, the PET remains in the as-heated shape.

This method was analyzed for use in origami applications [34] and has been demonstrated in a medical support system to mitigate buckling [35], lamina emergent origami [36], and in developable mechanisms [37]. This simple plastic annealing process can be very useful as it can change the energy behavior of the material, allowing the user to select different stable and unstable states [38].

2.3. Constant Force Springs

The self-retraction of the SPIRA Coil actuator is enabled by a material structure that closely resembles that of a "constant force spring". Votta's equation for the retraction force, F_R, of a constant force spring of natural diameter D_n is

$$F_R = \frac{Ebt^3}{26.4(\frac{D_n}{2})^2} \tag{1}$$

where E is the Young's modulus of the material, b is the spring's width, and t is the spring's thickness [39].

3. Design

The actuator is made using readily-available materials. The walls of the actuator's inflatable chamber are made of metalized PET. These sheets also provide the conductive path for the integrated resistance sensing. The exterior of the actuator is made from PET laminate material (heat-activated adhesive backed with PET), which bonds the actuator together and provides the pneumatic seal. A lay-flat PET tube is inserted at the open end of the actuator to allow air to enter once sealed, and shims are placed on the sides to help the edges properly adhere.

All of these layers are then rolled up on a mandrel. During the rolling process the layers are allowed to shift end-to-end, which prevents the wrinkling and local buckling often observed in coiled booms (see Section 2.1). Once rolled up and secured, the layers are then heat-treated. The heat treatment melts the heat-activated adhesive which bonds and seals the actuator together. The heat also serves to plastically anneal the PET material, relaxing the internal stresses. Once cooled, the actuator remains in the coiled shape.

After heat treatment, the shims and mandrel are removed and the formed actuator acts as a coiled PET spring with an integrated pneumatic chamber. As air enters the pneumatic chamber, the spring force is overcome, and the actuator unrolls and provides structure. As air is removed, the spring force rolls the actuator back up. In other words, the extension force comes from the air pressure, while the retraction force comes from its spring-like behavior.

The conductive aluminum of the metalized PET forms the circuit used in resistance sensing, as shown in Figure 2. During the extension and retraction, the metalized PET separates in the inflated portion and connects again at the coil (where the aluminum on one side of the chamber contacts the aluminum on the other side of the chamber). Hence, the resistance of the separated aluminum corresponds to the inflated length of the actuator.

3.1. Construction

The construction of each actuator is shown in Figure 3 and described below. A video of how to fabricate a SPIRA Coil actuator can be found in the Supplementary Materials as Video S2 (the video has minor changes to the steps below).

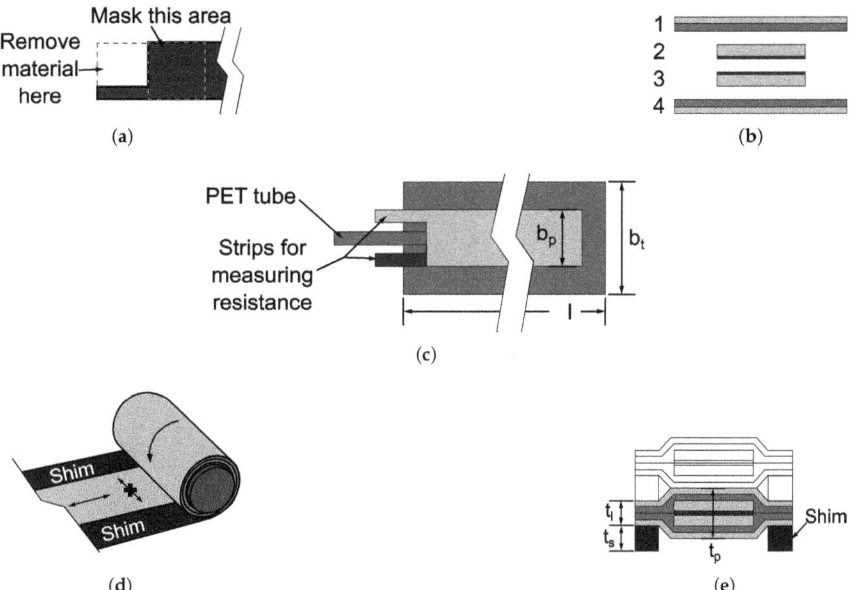

Figure 3. Steps in the manufacture of an actuator: (**a**) cut and mask the metalized PET (**b**,**c**) layer as shown; (**c**) insert PET tube (top laminate layer not shown); (**d**) place shim material, roll up, secure, and heat-treat in oven. Subfigure (**e**) shows a cross-section of two layers of a rolled-up actuator after heat treatment (not to scale).

First, two metalized PET layers are cut to width b_p and length l. These layers should be cut and masked as shown in Figure 3a, which allows the conductive layer to be accessible outside of the actuator without allowing the pressurized air to escape. The masking (placed on the conductive side of the material) prevents contact between the aluminum layers at the base of the actuator, as this area may become partially pinched during actuation.

Second, two strips of PET laminate are cut to width b_t and length l and the four strips are stacked together as follows (see Figure 3b):

1. PET laminate (adhesive face down);
2. Metalized PET (aluminum face down);
3. Metalized PET (aluminum face up);
4. PET laminate (adhesive face up).

Layers 2 and 3 should be shifted such that the strips for measuring resistance extend beyond the PET laminate as shown in Figure 3c. A laid-flat PET tube is inserted between layers 2 and 3, which will allow air to enter the sealed chamber in the completed actuator. A few centimeters of the opposite end can be adhered to prevent those ends from sliding during the rolling process.

Due of the difference in thicknesses between the laminate-only and pneumatic sections of the final actuator, a flexible shim must be chosen such that $t_p \approx t_l + t_s$. These disposable shims are placed over the laminate-only layers (see Figure 3e) during the fabrication process. This ensures that an adequate adhesive bond is made throughout the actuator.

Next, the layers are rolled up on a mandrel of diameter D_i, as shown in Figure 3d. Care should be taken that the layers do not shift side-to-side during the rolling process and to ensure that the electrical circuit is not compromised. Excess laminate material will accumulate at the unrolled end and can be trimmed. Secure the completed roll and place in an oven at approximately 160 °C until all the adhesive has bonded. The shims should be removed after the actuator has cooled.

For this paper, we used the materials found in Table 1.

Table 1. Materials used in this paper.

Material	Thick.	Vendor	Part No.
Metalized PET	2 mil	McMaster-Carr	7538T11
PET Laminate	5 mil	USI-Laminate	1146
Masking Polyimide	1.5 mil	McMaster-Carr	1754N13
PET Shim	5 mil	Various	Various

4. Model

The presented actuator can be modeled as a constant force spring and as a pneumatic cylinder. Hence, the force model can be viewed in two interlocking pieces: the spring model (for retraction), and the pneumatic model (for extension). Resistance sensing is also modeled.

4.1. Retraction: Spring Model

The spring model is based on Equation (1), with adjustments for different sections of our actuator. The actuator can be modeled as a constant force spring with two sections—pneumatic and laminate-only. The spring forces for these sections add in parallel and have the same diameter. The modulus of elasticity E, width b, and thickness t vary between these sections and will be noted by subscripts p and l for pneumatic and laminate-only, respectively, as shown in Figure 3.

If we assume that the diameter of the actuator is constant along its length ($D = D_i$), then the retraction force of the actuator, F_R, is

$$F_R \approx \frac{E_l(b_t - b_p)t_l^3 + 2(E_p b_p(\frac{t_p}{2})^3)}{26.4(\frac{D_i}{2})^2} \tag{2}$$

which considers the force from the pneumatic section to be the force of two identical springs of thickness $\frac{t_p}{2}$ in parallel.

For actuators in a coiled natural shape, the diameter, D, changes along the length x of the coil due to the thickness of the material. One way to approximate the diameter of a wrapped coil of a given length l is to calculate the cross-sectional area of the coil [40]. Since the cross-sectional area remains the same between coiled and uncoiled:

$$lt_p = \pi(\frac{D_o}{2})^2 - \pi(\frac{D_i}{2})^2 \tag{3}$$

where l and t_p are the length and thickness of the coiled sheet, and D_o and D_i are the outer and inner diameters of the coil. Solving for the outer diameter, D_o, we obtain

$$D_o = 2\sqrt{\frac{lt_p}{\pi} + (\frac{D_i}{2})^2} \tag{4}$$

We can also solve for the coil diameter D at intermediate states, where some length of material x is uncoiled and the rest of the material, $l - x$, is coiled:

$$D = 2\sqrt{\frac{(l-x)t_p}{\pi} + (\frac{D_i}{2})^2} \tag{5}$$

Combining Equations (2) and (5), we obtain an equation for the retraction force of our actuator, F_R, as a function of the deployed length:

$$F_R = \frac{E_l(b_t - b_p)t_l^3 + 2(E_p b_p(\frac{t_p}{2})^3)}{26.4(\frac{t_p(l-x)}{\pi} + (\frac{D_i}{2})^2)} \tag{6}$$

4.2. Extension: Pneumatic Model

Neglecting losses and elastic energy storage, the pneumatic extension force F_E is the product of the internal gauge pressure P and the change in volume per unit length $\frac{dV}{dx}$

$$F_E = P\frac{dV}{dx}. \tag{7}$$

The change in volume per unit length $\frac{dV}{dx}$ is equal to the cross-sectional area of the inflated portion of the actuator A_{pneu}

$$F_E = PA_{\text{pneu}}. \tag{8}$$

Except for the portion near each end, we assume that the pressure will force the cross-section the inflated portion to be circular, thus maximizing the volume within the constrained perimeter length. If we assume the cross-section is circular, then we can replace A_{pneu} with an expression based on circumference of the circle being equal to twice the width of the inflatable section, b_p

$$F_E = P\frac{{b_p}^2}{\pi}. \tag{9}$$

Because our actuator has a constant cross-section, this result matches with those obtained by [27,28].

4.3. Sensing: Resistance Model

The self-sensing of the actuator is performed by measuring the resistance of the vacuum-deposited aluminum that "metalizes" the inner PET layers. This resistance value can then be correlated with the length of the actuator.

The resistance of a wire can be found using the following equation:

$$R_{\text{wire}} = \frac{\rho_{\text{wire}} l_{\text{wire}}}{A_{\text{wire}}} \tag{10}$$

where ρ_{wire} is the resistivity, l_{wire} is the length, and A_{wire} is the cross-sectional area of the wire. For our application, the wire is the metalized PET strips. Assuming the vacuum-deposited aluminum has a constant thickness t_{al}, the resistance is a linear function of the length of the inflated portion of the actuator, x:

$$\Delta R = \frac{2\rho_{\text{al}} x}{b_p t_{\text{al}}} \tag{11}$$

In the case that the actuator completely extends, the metalized PET strips may completely separate and no electrical connection is made. This makes it easy to measure when the actuator is fully extended. The connection will be restored when the actuator starts to roll back up into an intermediate state.

5. Testing

5.1. Test Setup

The testing was primarily performed using a tensile test machine and custom fixture plates, as shown in Figure 4. The base of each actuator was clamped in the bottom fixture, which has a channel for the pneumatic inlet tube and copper contacts for electrical connection with the resistance sensor. The top fixture was constructed using a polished steel shaft which rolls in ultra-low-friction PTFE dry-running sleeve bearings and is covered by 12.7 mm (1/2") PTFE tubing. This allows the length of the actuator to be controlled and measured with the rolling portion of the actuator free to roll and slide on the rod with minimal friction. The force data found by the tensile test machine (F_M) is the sum of the pneumatic force (F_E) and the spring force (F_R), which include frictional losses.

The tensile test machine fully extended and retracted the actuator several times while collecting corresponding force data. Meanwhile, a simple pneumatic system was controlling the air pressure inside the actuator. A video of one cycle of a test is included in the Supplementary Materials as Video S3.

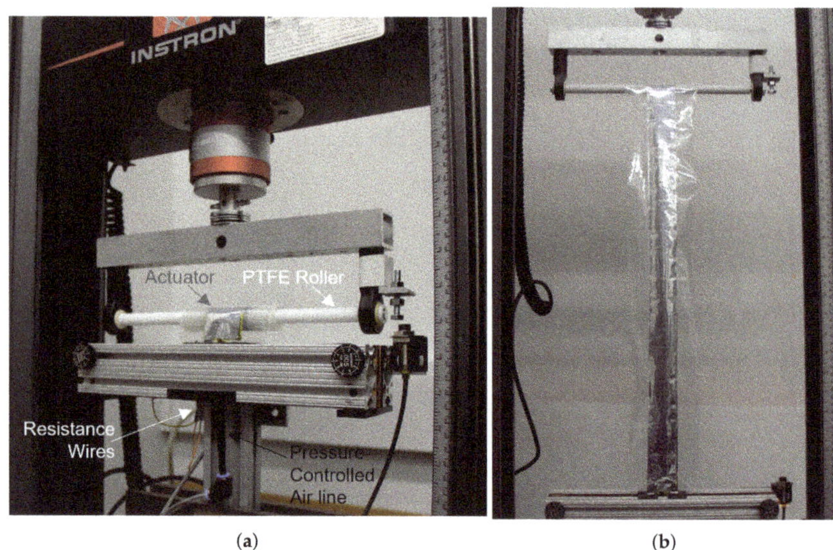

(**a**) (**b**)

Figure 4. (**a**) A tensile test machine was used to test various actuators under tightly controlled internal pressure conditions. (**b**) The actuators were tested at lengths up to 60 cm from their relaxed length.

5.2. Material Characterization

The effective modulus of elasticity for each of two sections of the actuator in Equation (2) was characterized based on sample measurements performed on the tensile test machine. Coils of laminate–laminate and laminate–metalized PET were prepared and tested on using the same setup as described above. From this data we estimated the modulus of elasticity values as approximately $E_p = 5\,\text{GPa}$ and $E_l = 3\,\text{GPa}$.

All the actuators shown were made using the same batch of metalized PET. We assume the aluminum has a constant thickness throughout the batch and that the aluminum has a resistivity of $\rho_{\text{al}} = 2.65 \times 10^{-8}\,\text{Ohm} \cdot \text{m}$. The resistance of a 2-cm-wide, 60-cm-long sample strip of metalized PET was measured to be 37.07 Ohms. Using Equation (10), the approximate thickness of the aluminum was calculated as $t_{\text{al}} = 1.87 \times 10^{-8}\,\text{m}$.

5.3. Spring Force Response

The spring response of the actuators was tested using the test setup described above with the tube inlet left open (Figure 5). The open tube ensured zero gauge pressure in the pneumatic section of the actuator ($F_E = 0$). Hence, the force measured by the tensile test machine is the spring force ($F_M = F_R$), which is considered to include hysteresis-inducing losses.

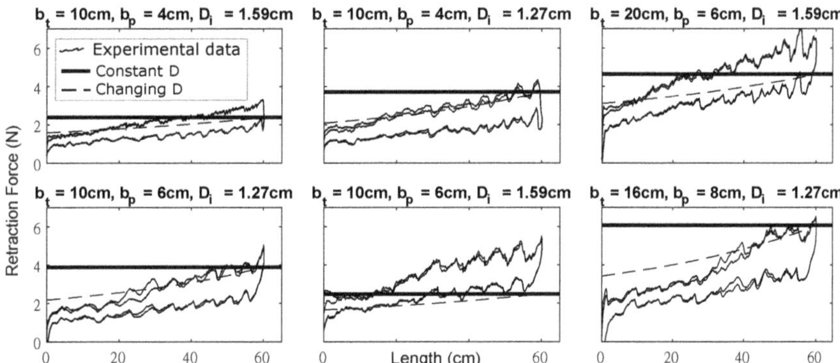

Figure 5. Measured spring retraction force compared to the model-predicted forces presented in Section 4.1 for six different actuators.

5.4. *Pneumatic Force Response*

The pneumatic force response testing was similar to the spring response testing but involved a constant applied pressure to the actuators' pneumatic input. A simple pressure controller with a large pneumatic accumulator was used to maintain constant pressure in the actuator during tensile testing. In order to keep the coil end of the actuator in consistent contact with the rolling shaft, the magnitude of the pressure was limited to those pressures which allowed the actuator to maintain tension on the test fixture in both extension and retraction. If a larger pressure was used, then fixture would apply compressive forces to the actuator, confounding the results and often causing the actuator to buckle and/or slip out of the fixture. This not only limited the number of distinct pressures that individual actuators could be tested at but limited the number of actuators that could be tested at multiple pressures.

5.5. *Resistance Sensing*

To achieve an accurate resistance measurement, a sufficient amount of pressure is needed to ensure that the metalized PET strips remain separated during actuation. In free extension and retraction, the actuator maintains sufficient back pressure for resistance sensing (for more details, see Section 6.3). It is also important to provide sufficient masking on the metalized PET, as described in Section 3.1.

The resistance was measured using a simple DAQ and voltage divider simultaneous with the pressure-response tests described earlier.

6. Results and Discussion

6.1. *Spring Force Response*

As can be seen in Figure 5, the model-predicted force values overlapped with the experimentally measured values. In almost every case, using the more complex, changing-diameter model (Equation (6), dashed line) resulted in a more accurate spring-force prediction than the simpler, constant-diameter model (Equation (2), thick solid line). The model, however, does not account for the evident hysteresis in the response.

6.2. *Pneumatic Force Response*

The results, shown in Figure 6, show the pneumatic force of several actuators ($F_E = F_M - F_R$) compared to the model predictions. The plotted experimental values have the spring component of the force (from Figure 5) subtracted from them to isolate the net effect of the pneumatic pressure. The thick, dashed lines show the extension force predicted by the simple pneumatic model, Equation (9).

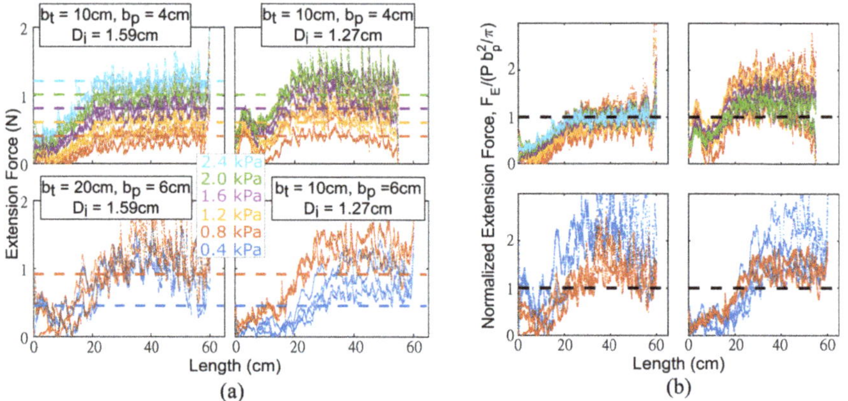

Figure 6. Extension force compared to the model prediction of Equation (9) for four different actuators. (**a**) The extension force plotted in these axes is defined as the measured tensile force minus the zero-pressure spring-retraction force for the same actuator. (**b**) These axes show the forces from (**a**) normalized by the model-predicted value (dashed lines).

At short extension lengths, the pneumatic force of the actuators starts small but, once a certain length is reached, the force becomes roughly constant, consistent with the constant force predicted by the model.

We again note that, to maintain tension in the fixture, the pressures shown in Figure 6 are lower than the self-extension of the actuators in unloaded conditions. The larger group of actuators shown in Figure 5 were found to have self-extension pressures that varied between 4 kPa and 9 kPa (0.6 psig to 1.3 psig).

Understanding the pneumatic force response of the actuator is an important step in the actuator design (e.g., to ensure equipment has sufficient pressures to actuate the device). In practice, once the minimum self-actuation pressure is met, the actuator extension is controlled more by mass than pressure. In the absence of external loads, the pressure inside the pneumatic chamber will equalize to the corresponding spring retraction force. Higher pressures do not always lead to higher extension forces. If the extension is constrained, the actuator may buckle or fold as internal pressure is increased. Hence, while the retraction force is controllable up to the maximum of the spring retraction force, the extension force is limited.

6.3. Resistance Sensing

Only one actuator prototype was free of manufacturing flaws and had sufficient retraction force to allow the resistance response to be reliably characterized on the tensile testing machine. As discussed previously, the pressure values at which we could test each actuator were limited. Some of these pressures were not large enough to fully separate the metalized PET layers along the entire length of the deployed actuator, creating an unusable resistance response at those pressures.

While this might seem like a problem, such low-pressure conditions are unlikely to be experienced by the actuator without active mass removal (i.e., vacuum). Pressurized self-extension produces sufficient back pressure for accurate resistance sensing. The same is true for self-retraction, where the back pressure is created by the flow restriction at the inlet. The experiments reported here included pressures that were lower than the pressures the interior volume would experience in most applications.

Further, several actuators had manufacturing flaws that affected the metalized PET. These flaws included the following: insufficient masking, inadvertent scoring and removal of aluminum along width of actuator, and buckling along wrinkle in laminate, allowing the PET strips to remain in contact in that area.

The resulting length to resistance data are shown in Figure 7. At extremely low pressures (i.e., below approximately 1.4 kPa or 0.2 psig), the resistance value was inconsistent due to insufficient back pressure, as discussed above. Above these pressures, however, the resistance response was very consistent and followed a similar trend to the linear response predicted by Equation (11) (black dashed line).

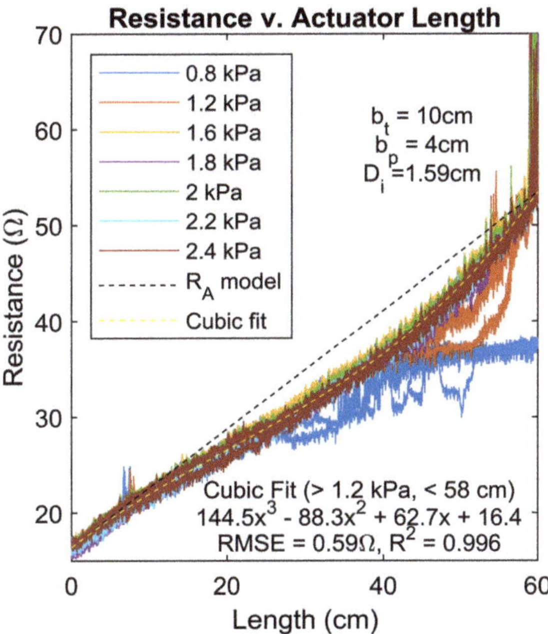

Figure 7. When a SPIRA Coil has sufficient pneumatic back pressure (in this case, ≈1.4 kPa or 0.2 psig), whether extending or retracting, the measured resistance can be used to determine the actuator length.

At full-extension, the calculated resistance became very high as the two sheets of metalized PET lost contact completely (nearly vertical lines in the top right of axes in Figure 7). This feature makes it very easy to identify the full-extension state of the actuator.

The measured resistance response at the pressures higher than 1.2 kPa and lengths less than 58 cm were fit with a cubic polynomial (yellow dashed line). The consistent response, virtually free from observable hysteresis allowed the cubic fit to describe nearly all the variance in the data, with an R^2 value of 0.996 and a corresponding RMSE of only 0.59 Ohms. This suggests that the measured resistance can reliably be used to calculate the length of the actuator for logging and control. Since the metalized layers are inside the sealed internal volume, they are protected from outside dust and debris.

7. Feedback Control

The predictable resistance response suggests that the measured resistance can reliably be used to calculate the length of the actuator for logging and control. This section demonstrates feedback control of SPIRA actuators using the integrated resistance sensing.

7.1. Feedback Test Setup

A vertical test fixture was created to test the use of the resistance in the feedback control of the actuators' length (Figure 8). The actuators were each connected to two small wooden mounting plates with thin layers of foam and strips of copper tape sandwiched between the plates and the actuator. The foam and copper ensured a consistent and reliable connection to the metalized PET at the ends of the actuator. The actuator was clamped between these mounting plates and wires were attached to the copper tape as shown in Figure 8.

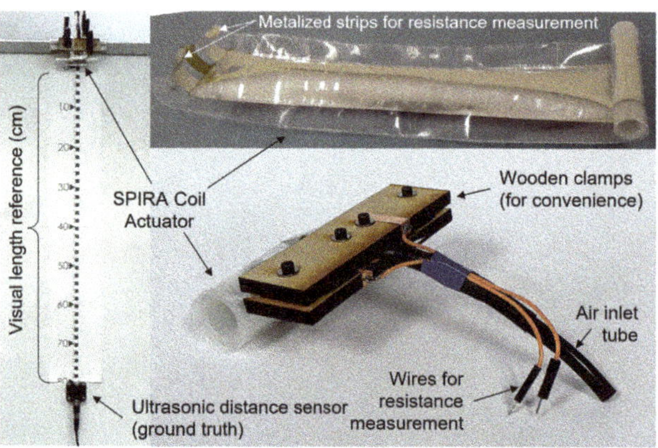

Figure 8. The SPIRA Coil actuator prototypes tested in Section 7.1 were outfitted with wooden clamps (for convenience), wires for resistance measurement, and an air inlet tube. Each actuator was connected to a vertical fixture for feedback testing using their internal-resistance-based length sensing. The vertical fixture included a printed ruler for visual length reference and an ultrasonic distance sensor for ground truth.

Next, the actuators were connected to a simple pneumatic system for adding and removing air, as well as an electronics system for measuring the actuators' resistance response. The actuator length in the fixture was measured by an ultrasonic distance sensor (DFRobot URM14, 1 mm accuracy).

A best-fit curve (second-order polynomial) relating the measured resistance to the actuator length (as measured by the ultrasonic sensor) was found for each actuator prior to feedback testing. The polynomial was used by the feedback controller to convert measured resistance values into estimated actuator lengths.

The actuators were then controlled using a simple proportional control algorithm, with the resistance providing the length estimate for feedback. Gain and dead-band values were chosen manually. The error in the commanded length was multiplied by a gain and used to drive the speed of a corresponding diaphragm pump. Two pumps and two solenoid valves were used, with one pair for inflation and the other for deflation. The nominal maximum flow rate of each pump was 12 L per minute (LPM) with a nominal power of 12 W.

The ultrasonic sensor was not included in the feedback loop and provided the ground truth for comparison. The sensor was specified to 1mm accuracy and ±1% error. However, due to its inherent limitations, the ultrasonic sensor was only able to capture distance data every 500 ms. The distance measurements happened during the first 100 ms of this window but were only reported by the sensor about 200 ms later. This may explain the slight time delay shown in the data. Spurious ultrasonic measurements were excluded from the reported data.

7.2. Feedback Results

The results from the feedback testing can be found in Figure 9, in a video included in the Supplementary Materials as Video S1, and in Table 2. As reported in Table 2, inspecting the measurement error between the feedback controller's resistance-based length estimate and the ultrasonic transducer measurement during the last five seconds of each commanded step, actuators 'a', 'b', and 'c' had steady-state respective length estimate errors of only 1%, 3%, and 2.3% on average, compared to the fully extended length of the actuators

(60 cm). For every commanded step across the three actuators, the average steady-state measurement error was less than 3 cm (5%).

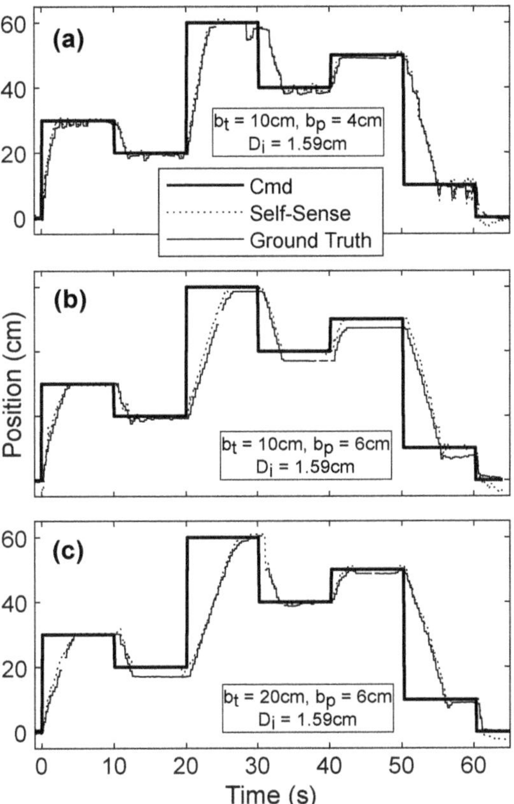

Figure 9. Time vs. distance graphs for three actuators (graphs 'a', 'b', and 'c' correspond to the three respective actuators) under feedback control. Internal metalized strips were used for the self-sense length feedback. Values from the ultrasonic sensor were used as the ground truth for comparison (and were not included in the feedback loop). Quantitative analysis is found in Table 2.

Table 2. A comparison of the measurement error (i.e., self-sensing value minus ground-truth) for the three actuators tested with feedback control during the last five seconds of each commanded step. Italicized values are in centimeters. The calculated percentage is based on the fully extended length of the actuator, 60 cm. Actuators 'a', 'b', and 'c' correspond to the respective actuators in Figure 9.

	Measurement Error at Each Commanded Step, cm						
Act.	**30 cm**	**20 cm**	**60 cm**	**40 cm**	**50 cm**	**10 cm**	**Average**
(a)	*0.46* (0.76%)	*0.67* (1.11%)	*0.03* (0.05%)	*0.97* (1.61%)	*0.71* (1.18%)	*0.79* (1.32%)	*0.60* (1.01%)
(b)	*−0.30* (−0.50%)	*1.00* (1.67%)	*1.71* (2.84%)	*2.74* (4.57%)	*2.86* (4.77%)	*2.82* (4.70%)	*1.81* (3.01%)
(c)	*0.22* (0.36%)	*2.65* (4.42%)	*1.89* (3.15%)	*0.74* (1.23%)	*1.08* (1.81%)	*1.51* (2.51%)	*1.35* (2.25%)

At some positions, the actuators had difficulty maintaining the commanded position and instead oscillated around the commanded position. This can be best seen in actuator

(a) of Figure 9. This may be caused in part to inherent mechanical hysteresis and the jagged force response (local mountains and valleys), shown in Figure 5, and relatively large pumps that were used in the control. Actuator (a), in particular, had a small air leak that acted as a disturbance, causing it to slowly retract when the pumps were stopped. As can be seen, the resistance measurement allowed the control system to identify and correct the effect of this disturbance.

For some of the feedback testing, control resulted in a steady-state error of a few centimeters (Figure 9). In these cases, it is clear that the controller believed the actuator was at the correct length (convergence of the dotted line with the thick black line) even as the ultrasonic sensor reported the error for post-processing (gap between the thin solid line and the thick solid line). This may be due in part to the best-fit calibration curve not fully capturing the resistance response.

8. Conclusions

The SPIRA Coil actuator presented in this work has many unique features compared to other actuators, including long extension lengths, ease of fabrication, low actuation pressure, and self-sensing. The coiled structure of the actuator allows it to extend to dramatically long lengths from a relatively small initial diameter. The actuator can be fabricated with low-cost, readily available materials and tools. The material for each actuator costs only a few US dollars. The fabrication requires little more than cutting tools and an oven. Compared to elastomer-based actuators, which often require large pressures to strain the material (e.g., 345 kPa for one of the most widely-cited soft actuators [41]), some of the SPIRA Coil prototypes tested in this work were able to self-extend with pressures lower than 5 kPa (less than 0.75 psig).

The simple spring and pneumatic models presented and validated in this work successfully approximated the experimental values. These simple-to-use algebraic models will enable SPIRA Coils to be readily designed for use in a variety of robotic applications, without the need to resort to complex and expensive computational modeling techniques.

Furthermore, the SPIRA Coils can measure their own extension through the resistance of the circuit formed by the metalized sheets, creating the opportunity for feedback control. Feedback control was successfully demonstrated on three actuators with an average measurement error of 1.2 cm (2.1%).

Compared to other potential electrical values (e.g., inductance, capacitance) resistance is by far the simplest to measure. Many implementations of resistance sensing in soft robotics, however, exhibit extreme time dependence and hysteresis due to the interaction of the resistive elements with elastomeric structures and the associated stress–relaxation [12]. In contrast, the resistance of SPIRA coils shows virtually no hysteresis (Figure 7). Though the resistance depends on the contact of two conductive surfaces, these surfaces are inside the sealed internal volume and thus are inherently protected from outside dust and debris which could foul that connection. Future work on SPIRA Coil self-sensing could include methods for preventing localized buckling along creases, preventing electrical connection along the edges, and an investigation into the effect of temperature on the resistance.

Compared to other soft actuators, SPIRA Coils can extend to dramatic lengths. They are low-cost, easy-to-fabricate, and easy-to-use. Their motion can be driven with low-pressure micro-pumps and their length can measured through resistance. The techniques and results presented in this work represent the beginning of the investigation into SPIRA Coils. We welcome additional work to improve and extend these methods. We are confident that the robotics community will find many interesting and useful applications for this unique actuator technology.

9. Patents

Felt, Wyatt and Greenwood, Jacob R. U.S. Provisional Appl. #63/489,962. *Self-Sensing Pressure-Driven Extending Actuator*. Filed: 13 March 2023.

Supplementary Materials: The following supporting information is available: Video S1: Video abstract and SPIRA Coil feedback control demonstration, https://youtu.be/fdyJlJAw4ao; Video S2: How to fabricate SPIRA Coils, https://youtu.be/dUjNiWvFAJc; Video S3: Force testing of SPIRA Coils, https://youtu.be/Ylq9VbeD4l8.

Author Contributions: Conceptualization, W.F. and J.R.G.; methodology, W.F. and J.R.G.; software, J.R.G. and W.F.; validation, J.R.G. and W.F.; formal analysis, W.F.; investigation, J.R.G. and W.F.; resources, W.F.; data curation, W.F.; writing—original draft preparation, W.F. and J.R.G.; writing—review and editing, W.F.; visualization, J.R.G. and W.F.; supervision, W.F.; project administration, W.F.; funding acquisition, W.F. All authors have read and agreed to the published version of the manuscript.

Funding: This material is based upon work supported by the U.S. Department of Energy, Office of Science, Office of Energy Efficiency and Renewable Energy, Building Technologies Office, under Award Number DE-SC0022837. This report was prepared as an account of work sponsored by an agency of the United States Government. Neither the United States Government nor any agency thereof, nor any of their employees, makes any warranty, express or implied, or assumes any legal liability or responsibility for the accuracy, completeness, or usefulness of any information, apparatus, product, or process disclosed, or represents that its use would not infringe privately owned rights. Reference herein to any specific commercial product, process, or service by trade name, trademark, manufacturer, or otherwise does not necessarily constitute or imply its endorsement, recommendation, or favoring by the United States Government or any agency thereof. The views and opinions of authors expressed herein do not necessarily state or reflect those of the United States Government or any agency thereof.

Data Availability Statement: The data presented in this study are available on request from the corresponding author.

Acknowledgments: The authors gratefully acknowledge the feedback and support of their colleagues at VPI Technology, especially Eric Brooksby. The authors would also like to thank the University of Utah Materials Characterization Lab for the use of their tensile test machine. This paper adapts and expands material that was presented at the 2023 ASME International Design Engineering Technical Conferences and Mechanisms and Robotics Conference (IDETC/MR 2023) under the title "SPIRA Coils: Soft Self-Sensing Pneumatic Integrated Retractable Actuator Coils" [42]. The authors of the original conference paper match those of the present paper, Jacob R. Greenwood and Wyatt Felt. Copyright for that work is held by ASME and it is used here by permission.

Conflicts of Interest: The authors declare no conflict of interest. The funding sponsors had no role in the design of the study; in the collection, analyses, or interpretation of data; in the writing of the manuscript, or in the decision to publish the results.

Abbreviations

The following abbreviations and nomenclature are used in this manuscript:

ASME	American Society of Mechanical Engineers
SPIRA Coil	Soft Self-Sensing Pneumatic Integrated Retractable Actuator Coil
A_{pneu}	cross-sectional area of the inflated pneumatic chamber
b	width of a constant force spring
b_p	width of the flattened pneumatic chamber
b_t	total actuator width
D	natural diameter of the actuator at length x
D_i	inner diameter of the actuator
D_n	natural diameter of a constant force spring
D_o	outer diameter of the actuator
E	Young's modulus
E_l	Young's modulus of laminate-only section
E_p	Young's modulus of pneumatic chamber
F_M	force measured by the tensile test machine
F_R	retraction force
F_E	extension force
l	total actuator length

P	pressure in pneumatic chamber
ρ_{al}	resistivity of aluminum, 2.65×10^{-8} ohm m
ΔR	resistance of the actuator
t	thickness of a constant force spring
t_{al}	thickness of aluminum on the metalized PET
t_l	thickness of laminate section
t_p	thickness of the flattened pneumatic chamber
t_s	thickness of shim material
V	volume of air in pneumatic chamber
x	deployed actuator length

References

1. Xavier, M.S.; Tawk, C.D.; Zolfagharian, A.; Pinskier, J.; Howard, D.; Young, T.; Lai, J.; Harrison, S.M.; Yong, Y.K.; Bodaghi, M.; et al. Soft pneumatic actuators: A review of design, fabrication, modeling, sensing, control and applications. *IEEE Access* **2022**, *10*, 59442–59485. [CrossRef]
2. Pagoli, A.; Chapelle, F.; Corrales-Ramon, J.A.; Mezouar, Y.; Lapusta, Y. Review of soft fluidic actuators: Classification and materials modeling analysis. *Smart Mater. Struct.* **2021**, *31*, 013001. [CrossRef]
3. El-Atab, N.; Mishra, R.B.; Al-Modaf, F.; Joharji, L.; Alsharif, A.A.; Alamoudi, H.; Diaz, M.; Qaiser, N.; Hussain, M.M. Soft actuators for soft robotic applications: A review. *Adv. Intell. Syst.* **2020**, *2*, 2000128. [CrossRef]
4. Veale, A.J.; Xie, S.Q.; Anderson, I.A. Characterizing the Peano fluidic muscle and the effects of its geometry properties on its behavior. *Smart Mater. Struct.* **2016**, *25*, 065013. [CrossRef]
5. Daerden, F.; Lefeber, D. The concept and design of pleated pneumatic artificial muscles. *Int. J. Fluid Power* **2001**, *2*, 41–50. [CrossRef]
6. Joe, S.; Bernabei, F.; Beccai, L. A Review on Vacuum-Powered Fluidic Actuators in Soft Robotics. In *Rehabilitation of the Human Bone-Muscle System*; IntechOpen: London, UK, 2022.
7. Felt, W.; Robertson, M.A.; Paik, J. Modeling vacuum bellows soft pneumatic actuators with optimal mechanical performance. In Proceedings of the 2018 IEEE International Conference on Soft Robotics (RoboSoft), Livorno, Italy, 24–28 April 2018; IEEE: Piscataway, NJ, USA, 2018; pp. 534–540.
8. Li, S.; Vogt, D.M.; Rus, D.; Wood, R.J. Fluid-driven origami-inspired artificial muscles. *Proc. Natl. Acad. Sci. USA* **2017**, *114*, 13132–13137. [CrossRef] [PubMed]
9. Hawkes, E.W.; Blumenschein, L.H.; Greer, J.D.; Okamura, A.M. A soft robot that navigates its environment through growth. *Sci. Robot.* **2017**, *2*, eaan3028. [CrossRef]
10. Felt, W. An Inverting-Tube Clutching Contractile Soft Pneumatic Actuator. *arXiv* **2019**, arXiv:1903.02725.
11. Hammond, Z.M.; Usevitch, N.S.; Hawkes, E.W.; Follmer, S. Pneumatic reel actuator: Design, modeling, and implementation. In Proceedings of the 2017 IEEE International Conference on Robotics and Automation (ICRA), Singapore, 29 May–3 June 2017; IEEE: Piscataway, NJ, USA, 2017; pp. 626–633.
12. Felt, W. Sensing Methods for Soft Robotics. Ph.D. Thesis, University of Michigan, Ann Arbor, MI, USA, 2017.
13. Nomura, K.; Takeuchi, H.; Sato, M.; Ishii, H.; Uemura, M.; Akaboshi, T.; Tomikawa, M.; Hashizume, M.; Takanishi, A. Development of a Self-Propelled Actively Bendable Colonoscope Robot with a "Party Horn" Propulsion Mechanism. In Proceedings of the Abstracts of the International Conference on Advanced Mechatronics: Toward Evolutionary Fusion of IT and Mechatronics: ICAM 2015.6, Tokyo, Japan, 5–8 December 2015; The Japan Society of Mechanical Engineers: Tokyo, Japan, 2015; pp. 68–69.
14. Schenk, M.; Viquerat, A.D.; Seffen, K.A.; Guest, S.D. Review of inflatable booms for deployable space structures: Packing and rigidization. *J. Spacecr. Rocket.* **2014**, *51*, 762–778. [CrossRef]
15. Secheli, G.A. *Modelling of Packaging Folds in Metal–Polymer Laminates*; University of Surrey: Guildford, UK, 2018.
16. Fernandez, J.M. Low-Cost Gossamer Systems for Solar Sailing and Spacecraft Deorbiting Applications. Ph.D. Thesis, University of Surrey, Guildford, UK, 2014.
17. Grande, A.M. Preliminary Analysis of Innovative Inflatable Habitats for Mars Surface Missions. Master's Thesis, Politecnico di Milano, Milan, Italy, 2020.
18. Secheli, G.; Viquerat, A.; Aglietti, G.S. A model of packaging folds in thin metal–polymer laminates. *J. Appl. Mech.* **2017**, *84*, 101005. [CrossRef]
19. Kamaliya, P.K.; Upadhyay, S.H.; Mallikarachchi, H. Investigation of wrinkling behaviour in the creased thin-film laminates. *Int. J. Mech. Mater. Des.* **2021**, *17*, 899–913. [CrossRef]
20. Katsumata, N.; Fujii, R.; Natori, M.; Yamakawa, H. Models of Membrane Space Structures with Inflatable Tubes. *Trans. Jpn. Soc. Aeronaut. Space Sci. Aerosp. Technol. Jpn.* **2010**, *8*, Pc_59–Pc_65. [CrossRef] [PubMed]
21. Arya, M.; Lee, N.; Pellegrino, S. Crease-free biaxial packaging of thick membranes with slipping folds. *Int. J. Solids Struct.* **2017**, *108*, 24–39. [CrossRef]
22. Zirbel, S.A.; Lang, R.J.; Thomson, M.W.; Sigel, D.A.; Walkemeyer, P.E.; Trease, B.P.; Magleby, S.P.; Howell, L.L. Accommodating thickness in origami-based deployable arrays. *J. Mech. Des.* **2013**, *135*, 111005. [CrossRef]

23. Bolanos, D.; Varela, K.; Sargent, B.; Stephen, M.A.; Howell, L.L.; Magleby, S.P. Selecting and Optimizing Origami Flasher Pattern Configurations for Finite-Thickness Deployable Space Arrays. *J. Mech. Des.* **2023**, *145*, 023301. [CrossRef]
24. Fang, H.; Lou, M.; Hah, J. Deployment study of a self-rigidizable inflatable boom. *J. Spacecr. Rocket.* **2006**, *43*, 25–30. [CrossRef]
25. Wang, J.T. *Deployment Simulation Methods for Ultra-Lightweight Inflatable Structures*; National Aeronautics and Space Administration, Langley Research Center: Hampton, VA, USA, 2003.
26. Yuan, T.; Tang, L.; Liu, J. Dynamic modeling and analysis for inflatable mechanisms considering adhesion and rolling frictional contact. *Mech. Mach. Theory* **2023**, *184*, 105295. [CrossRef]
27. Fay, J.; Steele, C. Forces for rolling and asymmetric pinching of pressurized cylindrical tubes. *J. Spacecr. Rocket.* **1999**, *36*, 531–537. [CrossRef]
28. Steele, C.; Fay, J. Inflation of rolled tubes. In Proceedings of the IUTAM-IASS Symposium on Deployable Structures: Theory and Applications: Proceedings of the IUTAM Symposium, Cambridge, UK, 6–9 September 1998; Springer: Dordrecht, The Netherlands, 2000; pp. 393–403.
29. Sandy, C.; Cadogen, D. Inflatable structures for space applications. *Nasa Tech. Memo.* **2000**, *208577*, 90–97.
30. Chen, T.; Wen, H.; Jin, D.; Hu, H. New design and dynamic analysis for deploying rolled booms with thin wall. *J. Spacecr. Rocket.* **2016**, *53*, 225–230. [CrossRef]
31. Schenk, M.; Kerr, S.; Smyth, A.; Guest, S. Inflatable cylinders for deployable space structures. In Proceedings of the First Conference Transformables, Seville, Spain, 18–20 September 2013; Volume 9, pp. 1–6.
32. Viquerat, A.; Schenk, M.; Lappas, V.; Sanders, B. Functional and qualification testing of the InflateSail technology demonstrator. In Proceedings of the 2nd AIAA Spacecraft Structures Conference, Kissimmee, FL, USA, 5–9 January 2015; p. 1627.
33. Taber, R.E. Method of Making a Self-Coiling Sheet, US Patent 3,426,115, 4 February 1969.
34. Sargent, B.; Brown, N.; Jensen, B.D.; Magleby, S.P.; Pitt, W.G.; Howell, L.L. Heat set creases in polyethylene terephthalate (PET) sheets to enable origami-based applications. *Smart Mater. Struct.* **2019**, *28*, 115047. [CrossRef]
35. Sargent, B.; Butler, J.; Seymour, K.; Bailey, D.; Jensen, B.; Magleby, S.; Howell, L. An origami-based medical support system to mitigate flexible shaft buckling. *J. Mech. Robot.* **2020**, *12*, 041005. [CrossRef]
36. Klett, Y. Paleo: Plastically annealed lamina emergent origami. In Proceedings of the International Design Engineering Technical Conferences and Computers and Information in Engineering Conference, Quebec City, QC, Canada, 26–29 August 2018; American Society of Mechanical Engineers: New York, NY, USA, 2018; Volume 51814, p. V05BT07A062.
37. Hyatt, L.P.; Lytle, A.; Magleby, S.P.; Howell, L.L. Designing Developable Mechanisms From Flat Patterns. In Proceedings of the International Design Engineering Technical Conferences and Computers and Information in Engineering Conference, Online, 17–19 August 2020; American Society of Mechanical Engineers: New York, NY, USA, 2020; Volume 83990, p. V010T10A013.
38. Greenwood, J.; Avila, A.; Howell, L.; Magleby, S. Conceptualizing stable states in origami-based devices using an energy visualization approach. *J. Mech. Des.* **2020**, *142*, 093302. [CrossRef]
39. Votta, F., Jr. The theory and design of long-deflection constant-force spring elements. *Trans. Am. Soc. Mech. Eng.* **1952**, *74*, 439–447. [CrossRef]
40. Peressini, A.L. Comparing Solutions of the Paper Roll Problem. In Proceedings of the Metropolitan Mathematics Club of Chicago, Chigago, IL, USA, 6 February 2009. Available online: http://mtl.math.uiuc.edu/special_presentations/JoansPaperRollProblem.pdf (accessed on 28 October 2023).
41. Mosadegh, B.; Polygerinos, P.; Keplinger, C.; Wennstedt, S.; Shepherd, R.F.; Gupta, U.; Shim, J.; Bertoldi, K.; Walsh, C.J.; Whitesides, G.M. Pneumatic networks for soft robotics that actuate rapidly. *Adv. Funct. Mater.* **2014**, *24*, 2163–2170. [CrossRef]
42. Greenwood, J.R.; Felt, W. SPIRA Coils: Soft Self-Sensing Pneumatic Integrated Retractable Actuator Coils. In Proceedings of the 47th Mechanisms and Robotics Conference (MR), International Design Engineering Technical Conferences and Computers and Information in Engineering Conference, Boston, MA, USA, 20–23 August 2023; Volume 8.

 actuators

Article

Octopus-Inspired Robotic Arm Powered by Shape Memory Alloys (SMA)

Shubham Deshpande and Yara Almubarak *

SoRobotics Lab, Department of Mechanical Engineering, College of Engineering, Wayne State University (WSU), Detroit, MI 48208, USA; shubhamd@wayne.edu
* Correspondence: yaraalmubarak@wayne.edu

Abstract: Traditional rigid grippers that are used for underwater systems lack flexibility and have lower degrees of freedom. These systems might damage the underwater environment while conducting data acquisition and data sampling. Soft robotics, which is mainly focused on creating robots with extremely soft materials are more delicate for the grasping of objects underwater. These systems tend to damage the underwater ecosystem in the least possible way. In this paper, we have presented a simplified design of a soft arm inspired by the octopus arm actuated by coiled Shape Memory Alloys (SMAs) using completely flexible lightweight material. The characterization arm performance under various load and input current conditions is shown. We hope this work will serve as a basis for the future of underwater grasping utilizing soft robotics.

Keywords: coiled SMA; soft robot; biomimetic; artificial muscles; underwater grasping

Citation: Deshpande, S.; Almubarak, Y. Octopus-Inspired Robotic Arm Powered by Shape Memory Alloys (SMA). *Actuators* **2023**, *12*, 377. https://doi.org/10.3390/act12100377

Academic Editor: Hamed Rahimi Nohooji

Received: 16 August 2023
Revised: 24 September 2023
Accepted: 29 September 2023
Published: 4 October 2023

1. Introduction

Traditional robotic systems that are used for various applications such as medical, manufacturing, as well as exploration are always a bit invasive to the environment they are working in. This is due to the rigid body components of such systems. Even though rigid robotic systems get the job done, they can be inefficient, large, bulky, and expensive, making them unfavorable for delicate tasks. Work is being carried out to increase the efficiency of such systems. However, a better approach would be to take inspiration from a highly advanced system that has been evolving for billions of years: Nature.

Engineering problems can be approached by taking inspiration from nature's solution to various problems. This design, known as the biomimetic design approach, uses natural principles, structures, and processes as a basis for innovation and problem-solving in a wide range of fields. Soft robotics is another field that makes use of soft and flexible materials, such as silicone and elastomers, and allows the robotic system to adapt to its environment, conform to different shapes, and exhibit behaviors like living organisms. One of the key benefits of using biomimetic soft robots is that they can help to create more sustainable and efficient products and systems. A biomimetic system can use various actuation techniques depending on its application. Many such robotic systems make use of artificial muscles as their actuation technique and sensors. Artificial muscles are used to mimic the function of biological muscles when subjected to an external stimulus. Some examples of artificial muscles that can be used in biomimetic soft robots are Electroactive Polymers (EAPs) [1,2] that change their shape in response to an electric current, Pneumatic Artificial Muscles (PAMs) [3,4] that are made from elastomeric tubes which expand when they are pressurized, Twisted and Coiled Polymer muscles (TCPs) [5–7] and Shape Memory Alloys (SMAs) [8,9] that contract and expand in response to change in temperature. Each of the actuation techniques, including all the artificial muscles, has its advantages and disadvantages. A trade-off of one property while compensating for another property is necessary while developing any robotic system. Selecting an actuator that is compatible with the structure and can accomplish the designed tasks efficiently is critical to the success of the robotic system.

For underwater applications, although electroactive polymers and pneumatic artificial muscles work, their performance is greatly affected by the conductivity of the surrounding water medium. The performance and the actuation capacity of Shape Memory Alloys (SMAs) and Twisted and Coiled Polymers (TCPs) remain unaffected even if they are used in the underwater medium. While electrical motors and fluidic actuators are energy efficient and easily controlled, they tend to be bulky, expensive, and produce a lot of noise and disturbances in an underwater environment. Hence, using SMAs and TCP muscles is a great alternative to using traditional actuation techniques. On top of these advantages, artificial muscles such as TCPs and SMAs also have higher force density [10], higher strain [7,11,12], faster actuation [13], high endurance [5,14], and are lightweight.

The promising results of the Twisted and Coiled Polymer (TCPs) muscle behavior were shown by Haines et al. [15] and opened new ways in which these muscles can be created and modified. The behavior of these muscles embedded inside a soft silicone material was further investigated, and the optimized parameters were observed in [7]. Moreover, these muscles have demonstrated good grasping abilities in the remotely controlled octopus robot [6]. There has been research to investigate the inculcation of TCPs in devices for medical use. One such attempt is presented in [16], where a hand orthosis device has been fabricated and investigated. TCPs have also been used to accurately mimic the locomotive swimming nature and the appearances of a couple of jellyfish found in nature, namely, Chrysaora achlyos [12,14,17]. Another artificial muscle whose performance is unaffected due to the conductivity of the surrounding medium, specifically, in underwater environments is the Shape Memory Alloys that remember their memory shape and revert to it when the temperature increases.

Previously, SMAs have been employed in soft robots to mimic various aspects of the animals found in nature. The potential of linear Shape Memory Alloys coupled with silicone rubber tubes in a soft robotic actuator for gripping objects has been demonstrated [18]. A robotic structure that can swim in multiple directions using linear SMAs confined in a conduit and actuated by electrical current stimulation and mimics the appearance and biological locomotion of a jellyfish found in nature, Chrysaora Colorata is presented in [5]. Coiled SMAs have been used to create a soft robotic hand able to lift off weight up to 130 g through the process of Joule heating [19]. Coiled SMAs can be easily fabricated from linear SMAs. One such study presents the fabrication of coiled SMAs from linear SMAs and the characterization of the resulting coiled SMAs for using them in humanoid robots [10]. Hence, both coiled and linear SMAs have very high potential in applications for soft robots. Some advantages of using these artificial muscles in underwater soft robots are that they are flexible, noninvasive to marine life, durable, and typically have a low cost of operation. However, researchers are working on solving the challenges faced in developing these robots. Some of the challenges faced are the development of small, efficient, lightweight power sources and control systems, durability, and control due to the lack of rigidity in them, scaling up a soft robot also becomes a challenge as when the soft materials become larger, they tend to become less and less stable. Mimicking the anatomy and physicality of an octopus arm is a common solution for underwater manipulation due to its high degree of freedom. Varying octopus-inspired designs and actuation methods have been presented such as tendon-driven [20–22], actuated by SMAs and EAPs [2,23–26], hybrid actuation using stepper motors and fishing line TCP [6], pneumatics and fluidic [3,4,27–30]. Some insights regarding creating soft robots while having a minimum invasion of marine life are presented in the survey paper [31]. One of the first attempts at creating a biomimetic robot that moved on its own without requiring an external power supply was the Octobot [27], which is a 3D-printed soft-bodied autonomous robot. PoseiDRONE is a robot mimicking the maneuverability of an octopus through jet propulsion by the virtue of its equidistant arms around the body [21]. A dynamic model of a completely soft octopus-mimicking robot with an elastomeric main body has been presented in [32]. A bipedal walking of a soft robot inspired by a coconut octopus has been studied. The gait analysis of the bipedal octopus is presented along with the design of the mechanical structure and the control systems [33].

In this paper, a simplistic version of a biomimetic soft robotic octopus tentacle using coiled Shape Memory Alloys embedded in Ecoflex silicone is presented. The bending angle characteristics of the arm are studied when it is freely suspended with and without weights attached at the tip in an underwater environment and subjected to Joule heating (some examples are presented in supplementary movie S1). A simple temperature model is created that can be used to predict the temperature of an SMA when it is actuated under embedded conditions. The key highlights of this work can be summarized as follows:

- We introduce a modular and lightweight bio-inspired robotics arm. The design is simple and emphasizes flexibility utilizing artificial muscles as actuators.
- We develop a predictive temperature model for the coiled SMAs when actuated in silicone. This model aids in studying the control behavior of the SMAs during operation.
- Extensive actuation analyses reveal the arm's capabilities, including controllable 2D and 3D bending, a lifting capacity of up to 125 g, and the ability to grasp and hold objects of diverse sizes and shapes.

2. Materials and Methods

2.1. Design of the Silicone Arm with Embedded Coiled SMA

The proposed silicone arm shown in Figure 1 was created by embedding the coiled Shape Memory Alloys inside Ecoflex silicone with a shore hardness of 0010. The arm is in the shape of an inverted frustum of a cone with a base diameter of 50 mm and a truncated diameter of 20 mm with a silicone thickness of 3 mm. The total length of the arm is 200 mm. A coiled SMA with a spring wire diameter of 0.008″ (0.203 mm) and spring outer diameter of 0.054″ (1.37 mm) is crimped on both ends and embedded inside the silicone.

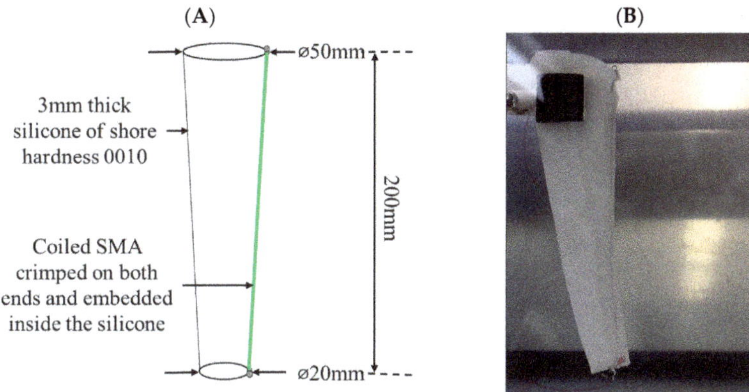

Figure 1. The design of the arm: (**A**) The conceptual design, (**B**) The fabricated silicone arm suspended underwater.

2.2. Fabrication Methods Used for Creating the Silicone Arm

The three-dimensional silicone arm is created by folding a trapezoidal silicone sheet into a truncated cone shape, as shown in Figure 1. To create the trapezoidal sheet, the CAD of the mold is designed in Fusion 360 and is 3D printed with PLA filament using Lulzbot Taz 5, as shown in Figure 2.

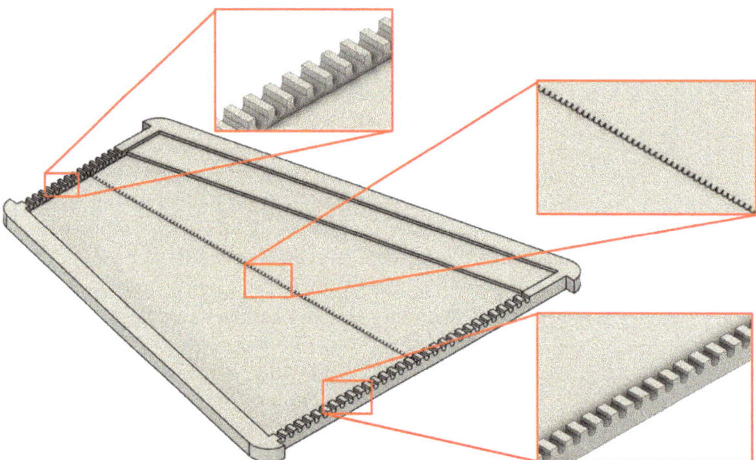

Figure 2. The CAD model of the PLA mold that was created in Fusion 360. The figure shows the schematic of the SMA coil and the slots present in the mold for fixing it.

The mold has multiple slots to fix the crimped SMA in place before embedding it in silicone. The slots are 2.5 mm in diameter and are centered 2.5 mm from the top surface of the mold. The silicone curing area of the mold is 3 mm thick, and the sides are equal to the circumference of the diameters of the arm. The smaller base of the trapezium is 60 mm long, while the larger base is 160 mm long. A lip of 20 mm is added at the side of the trapezium to seal the 2D sheet into a 3D truncated cone.

Firstly, the length of the SMA coil is chosen in such a way that when the coil is tensioned with a pull force of 0.089 pounds (38.3 g), the length becomes 200 mm. The compressed spring is hung vertically, and a calibrated weight of 38.3 g is attached to it. After tensioning the coil, the 3D-printed PLA mold is sprayed with a BK30 mold-release lubricant. The tensioned and crimped SMA is fixed in the mold, as shown in Figure 2. Next, 40 mL each of parts A and B of the platinum-cured Ecoflex 0010 were mixed and poured over the SMA fixed on the PLA mold. The silicone is set for curing at room temperature for 4–5 h. After the silicone is cured, the flat sheet of silicone that is embedded with the SMA is carefully removed from the mold, and the excessive silicone around the edges is cut. The 2D flat sheet is then folded over the sides, and the lip is sealed over the arm with small amounts of fast-curing Ecoflex 0035.

2.3. Experimental Setup of the Bending Characteristics of the Arm

The silicone arm is suspended in a tank underwater such that the SMA is exactly on the side of the arm, as shown in the schematic in Figure 3. The value of M for studying the bending characteristics is 0 g, as no weight is attached to the tip of the arm. The SMA embedded inside the arm is actuated with all combinations of 1.125 A–1.5 A with an interval of 0.125 A and heating times of 2 s–5 s with an interval of 1 s and the cooling time of 45 s underwater. All the actuation cycles were captured in a video recording.

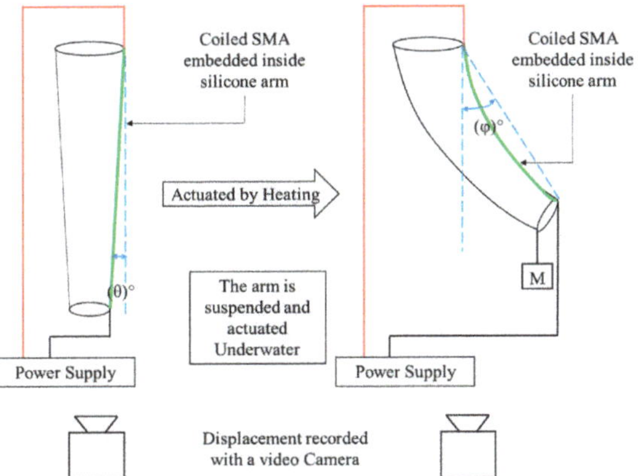

Figure 3. The schematic representation of the bending characteristics of the silicone arm embedded with coiled SMA.

The final videos of the actuation cycles were analyzed in Tracker Physics software. As shown in Figure 4A,B, a cartesian coordinate system was set up with the origin at the corner of the fixed stand. The final bending angle (plotted below) is given by the expression ($\varphi - \theta$), where φ and θ are as shown in the figure below. The bending angle of the arm with respect to the initial position as described in the mathematical description was studied as a function of the actuation current and actuation time.

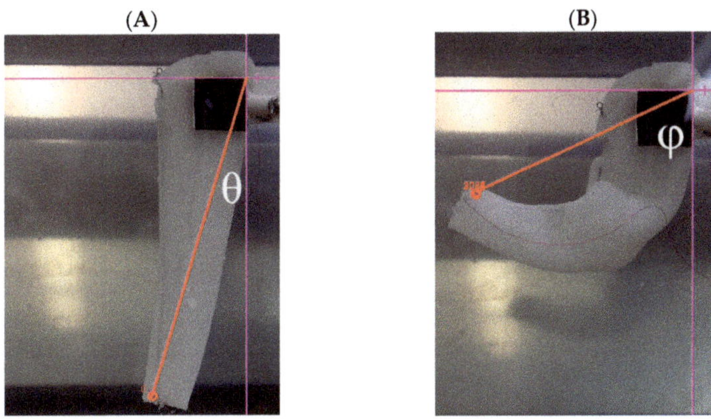

Figure 4. The placement of the origin of the cartesian coordinate system and the demonstration of the bending angle calculations. (**A**) Nondisplaced arm position. (**B**) At a fully actuated position.

2.4. Weightlifting Characterization of the Single SMA Embedded Inside the Silicone Arm

The experimental setup for performing the weightlifting capabilities of the SMA coil is shown in Figure 3, except that the mass M is nonzero in the case of the weightlifting characterization.

The broader end of the arm was hung with the help of a stand, as shown in Figure 5A. The other end of the arm was free to move. Figure 5B,C show the full displacement of the tip of the silicone arm when no weight is attached to the other end and with weight, respectively. The arm was actuated with the same combinations of the current and the heating times. Standard calibrated weights of 25 g, 50 g, and 75 g were used for this experiment. The bending angle of the arm with respect to the initial position when weights are attached with the help of a fishing line was recorded.

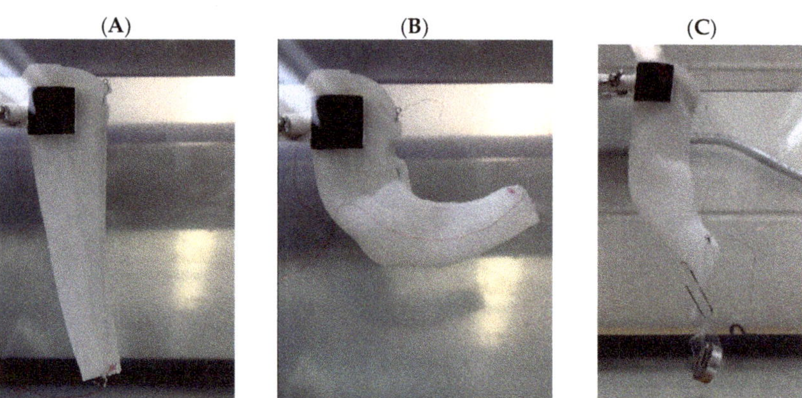

Figure 5. Some bending configurations of the silicone arm. (**A**) Nondisplaced arm. (**B**) Fully curled arm with no weight attached at the tip. (**C**) Fully displaced arm lifting 50 g.

2.5. Temperature Modeling of the SMA Embedded in the Silicone Arm

The coiled SMA embedded silicone arm is suspended underwater with the help of a lab stand. The J-type thermocouple wire which is connected to a NI-cDAQ 6178 chassis is inserted in the silicone and wound on the SMA coil in the geometrical center to obtain the temperature readings that are not affected by the extremities. The NI-cDAQ 6178 chassis reads the temperature of the coil with the help of LabVIEW. The frequency of the data acquisition is 10 Hz (explained in Figure 6). A simplistic analytical temperature model of the arm is developed to predict the temperature of the SMA coil. One can say from the first law of thermodynamics using Equation (1),

$$mC_p\frac{dT}{dt} = -hA(T - T_\infty) + I^2R \quad (1)$$

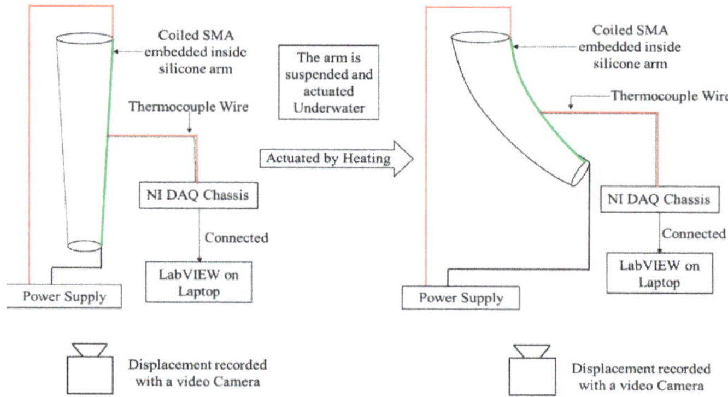

Figure 6. The schematic figure of the experimental setup for measuring the temperature of the embedded SMA coil.

Integrating this equation, we obtain Equation (2),

$$T = T_\infty + \frac{IV}{hA}\left[1 - exp\left(\frac{-hAt}{mC_p}\right)\right] \quad (2)$$

The values and the descriptions of each of the parameters in Equations (1) and (2) are given in Table 1.

Table 1. The parameters used for temperature modeling of the SMA coil embedded inside the silicone arm.

Quantity	Description	Value
$T_{a?\_}$	Water Temperature	17.5 °C
I	Current	1.125 A–1.5 A **
V	Voltage	18.9 A–23.0 A **
h	Convective Heat transfer coefficient	Obtained semi-analytically, partly through the experiments
A_s	The surface area of the coil exposed to the silicone and/or water	0.012 m^2
m	The mass of the SMA coil by itself	0.00228 kg
C_p	Specific heat capacity of the SMA + Silicone	1620 J/kg/°C
Duty Cycle	The ratio of the heating time to cooling time	4.2–10% **
Rate of change of temperature per unit time	Extracted Experimentally	12.12 °C/s ***

** These parameters are changed with every experiment depending on the used sets of actuations. *** This value is extracted from the heating cycle of the experimentally measured temperatures. All the heating cycles are considered (for all current and time combinations). The rate of change of the temperature over the heating cycle is calculated for all the experiments. The average value of the slopes is calculated and is found to be 12.12.

Equation (1) is a first-order differential equation containing the rate of change of temperature with respect to time and the temperature of the SMA coil at that specific time. To solve this equation, a Simulink model is created in MATLAB. The input to the Simulink model is the parameters as defined in Table 1. The experimental temperature output of LabVIEW in each case is recorded and plotted against time. In this section, the model was validated with the help of the experimental temperatures. The model was run alongside combinations of temperatures in which either only the current is varied or only the heating time is varied.

3. Results and Discussion

3.1. Bending Characterization of the Silicone Arm

Figure 7A–D show the bending angle of the tip of the arm vs. time for the currents used. As expected, a higher bending angle is caused by actuating the SMA with a higher current and a corresponding higher heating time. However, as can be seen in Figure 7B, the bending angle peaks at a minimum current of 1.25 A and a corresponding minimum time of 3 s. The maximum bending angle achieved is about 48°, as can be seen in Figure 7. This peak angle can be set as the limit of the silicone arm above which the arm is not able to bend.

For some heating time and current combinations (for example, 1.125 A for 2 s of heating), the tip of the arm does not reach the maximum bending angle. For most such cases, it can be said that the coil is not getting enough time to be heated, and the surrounding silicone and water are cooling the coil before it can reach the actuation temperature. For most higher current and heating time combinations, the displacement reaches the maximum value of displacement before the heating time ends. In such cases, as the tip has already reached maximum displacement, further supplying current to the coil heats the coil excessively. This excessive heat may damage the SMA coil or the silicone arm. If the SMA coil is repeatedly supplied with excessive heat while it is under tension, it may cause the coil to change its memory shape effect due to the martensite–austenite phase transformation. Such an alloy with a changed memory shape does not "remember" the original shape. Thus, as a result, the coil is no longer useful.

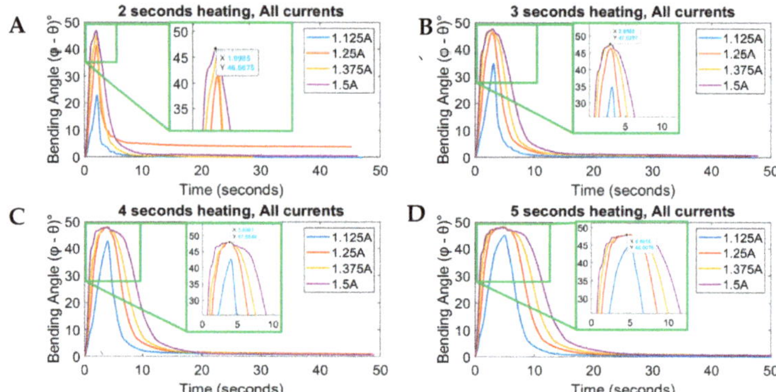

Figure 7. The plot shows the bending angle (in °) of the tip of the arm vs. the time (in seconds). (**A**) 2 s heating, (**B**) 3 s heating, (**C**) 4 s heating, (**D**) 5 s heating.

3.2. Weightlifting Characterization of the Arm with a Single Embedded Coiled SMA

In the weightlifting characterization experiment, the bending angle and the actuation voltage were recorded for all current and time combinations. The overall energy transferred to the SMA ($Q = VIt$) is calculated and tabulated below. The bending angle was plotted against the calculated energy for different loads that the arm was able to lift off the bottom of the fish tank.

From Figure 8, on average, the bending angle for the lower mass of 25 g is higher than the corresponding bending angle for the higher mass. The heat consumed, which is plotted on the X-axis, is a function of the voltage, the current, and the time. When the SMA coil is heated up to actuate it, the temperature of the coil increases due to Joule heating. This, in turn, reduces the resistance of the coil. As a result, the voltage across the coil is not constant but is constantly changing. For this experiment, the resistance is assumed to be constant throughout the experiment. Hence, although some points show deviation from one another in terms of the bending angle with respect to the energy consumed, the differences can be assumed to be in the experimental uncertainty.

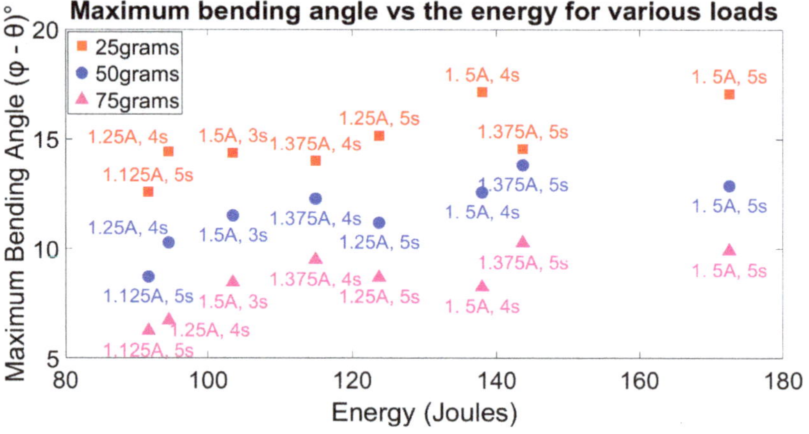

Figure 8. Maximum bending angle vs. energy consumed by the SMA for various loads.

It can be seen from the figure above that for low actuation current and low time, the bending angle of the tip of the arm does not reach its maximum potential. For a relatively lower load of 25 g on the arm, the maximum bending angle reaches 12.5°, whereas for a higher load of 75 g, the single-coil SMA arm can reach a maximum bending angle of 6° only. For all the loads and the current and time combinations, the bending angle peaks at around 130 J–140 J of electrical energy. At around 140 J, a maximum bending angle is achieved. This peak angle is around 17° for 25 g, 13° for 50 g, and about 10° for 75 g, as can be seen in Table 2. If a higher energy is supplied, by either increasing the current and/or the heating time of actuation, the bending angle saturates. However, the desired curling shape is achieved much faster than the curling shape with lower current and time combinations. The energy used is a trade-off between the frequency of actuation and the energy consumed during the experiment. If little to no energy is to be wasted in increasing the temperature of the coil excessively, then one has to trade off the faster frequency. However, if the arm is to be used in an application that requires higher frequency, then high energy can be supplied, and as a result, the measurable qualities such as efficiency and the overall life of the SMA are compensated.

Table 2. Tabulated values of the energy, voltage, and current are shown, which are plotted in Figure 8.

I	t	Power Density	V	Work = (VIt)	Maximum Bending Angle		
A	sec	kW/kg	Volts	Joules	25 g	50 g	75 g
1.125	5	8.04	16.3	91.7	12.58	8.73	6.25
1.25	4	10.36	18.9	94.5	14.43	10.29	6.72
1.5	3	15.13	23.0	103.5	14.38	11.52	8.47
1.375	4	12.60	20.9	114.9	14.01	12.28	9.50
1.25	5	10.86	19.8	123.8	15.18	11.19	8.69
1.5	4	15.13	23.0	138.0	17.17	12.58	8.26
1.375	5	12.61	20.9	143.7	14.57	13.81	10.28
1.5	5	15.13	23.0	172.5	17.09	12.88	9.93

Through the previous experiments, it was found that the input energy of 140 J allows the arm to reach the maximum bending angle during weightlifting. All energy supplied higher than 140 J goes into increasing the temperature of the coil without changing the shape of the silicone arm. Prolonged exposure to such high energy can cause the SMA coil to be damaged and change its original memory shape. Hence, the optimal energy to be supplied to the single-coil SMA embedded in a similar silicone arm is around 140 J, which can be achieved by passing 1.5 A for 4 s or 1.375 A for 5 s.

3.3. *Weightlifting Characterization of the Arm with Multiple Embedded Coiled SMAs*

Similar to the previous section, SMAs are embedded in the silicone; in this case 3 SMAs are embedded in parallel close to each other. The bending angle is recorded while carrying various loads. The arm shows good weightlifting capabilities when 3 SMAs are actuated together. This shows a linear trend between the number of SMAs embedded and the maximum weight it can carry. In this case we found that 3 SMAs, of the same coiled diameter, can carry up to a 125 g load with acceptable bending (~10°). The current and time combination used for the 3 SMAs is the same as for the single SMAs (multiplied by number of SMAs). Figure 9 shows the bending angle vs. load of 3 SMAs embedded vs. 1 SMA embedded. The results show a significant increase in the bending angle. We assume that increasing the number of SMAs embedded will further enhance the capability.

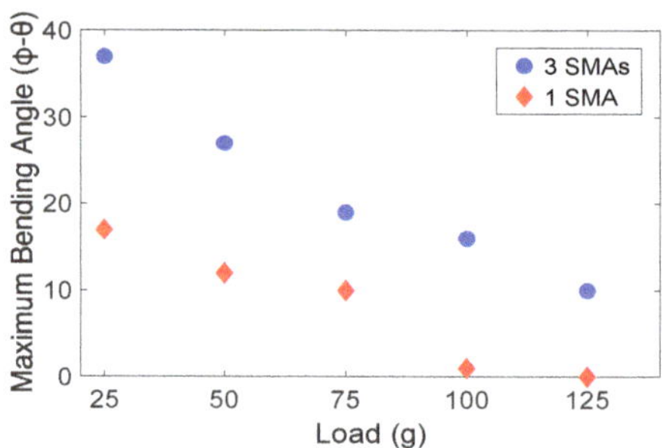

Figure 9. A comparison of the maximum bending angle vs. load for single and 3 SMAs embedded in the silicone arm.

3.4. Temperature Modeling of the Single SMA Coil Embedded in a Silicone Arm

The bending angle of the tip of the silicone arm is directly proportional to the temperature of the SMA coil. A low temperature of the actuated SMA results in a lower bending angle of about 20° for a combination of low current (1.125 A) for a low time (2 s). This is because the electrical energy supplied to the coil is not enough to raise the temperature of the coil to activate the shape memory effect. For most combinations of current and time, the temperature of the coil almost reaches the required value for the actuation to begin and activate the shape memory effect. Similar observations as section B can be made about these results as well, that if excessive energy is passed through the coil, then the resulting bending angle will not be more but will be limited at a certain value. All the energy transferred to the circuit after the limiting value will increase the temperature of the SMA coil and might destroy it.

Two types of temperature modeling results are shown in Figures 10 and 11, respectively. In Figure 10, the created model demonstrates the comparison between the experimental temperature and the analytical temperature obtained from the model presented in Equation (2) for an actuation current of 1.5 A for 2, 3, 4, and 5 s. In this test, the heating time is changed, which results in the change of the duty cycle, which is one of the crucial variables in the whole model. The other type of model demonstrates the comparison between the experimental temperature and the analytical temperature obtained from the model for a heating time of 5 s and a cooling time of 45 s for 1.25 A, 1.375 A, and 1.5 A of actuation current. In this type of model, the current is changed, which results in the change of voltage and the expression for h only.

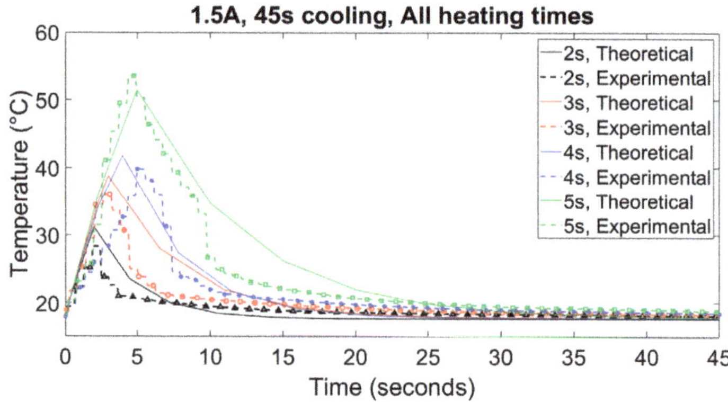

Figure 10. The experimental vs. theoretical temperature for an actuation current of 1.5 A for all the values of heating times.

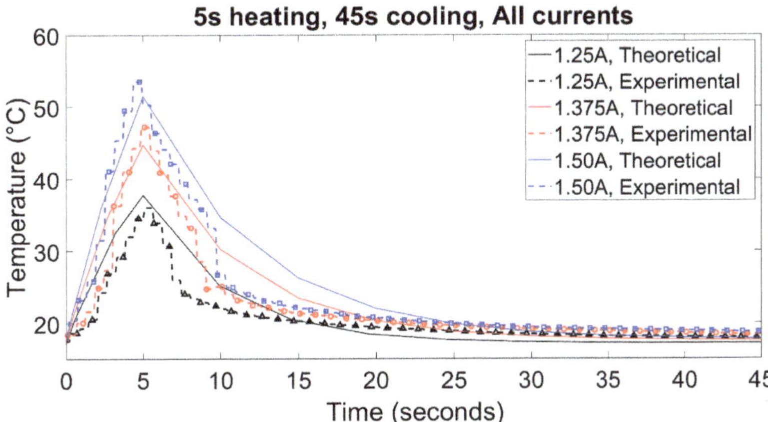

Figure 11. The experimental vs. theoretical temperature for a heating time of 5 s, for all the values of current.

Together, the results from these 2 models can be superimposed on one another, which results in a model that can predict the temperature of the SMA coil embedded inside the silicone arm when the current is changed and/or the heating time of the actuation cycle is changed. Thus, these 2 models were created individually and will be superimposed as and when needed. By performing this, a bigger model that predicts the temperature when both the current and the duty cycle changes can be created. Figure 10 shows the comparison between the experimental and analytical temperatures for 1.25 A, 1.375 A, and 1.5 A for a heating time of 5 s and a cooling time of 45 s.

As the model is designed for the heating cycle of the actuation only, for all the heating times, the model predicts a temperature that is closer to the experimental temperature. However, for the cooling cycle of the actuation, the model shows a difference between the experimental temperature and the model temperature. We can use these results to form a relation between the actuation parameters of the SMA and the arm bending, hence reducing experimental time and cost to develop a more robust soft arm for underwater application.

3.5. Grasping and Shape Manipulation of the Arm

Using the created arm, various objects with different sizes and weights were allowed to be grasped by the arm underwater. In these experiments, multiple SMAs were actuated. Figure 11 shows the various objects that the arm was able to lift underwater. It must be noted that the objects had to be held in place for a brief period while the actuation was activated for the arm to be able to hold them. In all the cases shown, the arm was actuated with 2.75 A (total current) for 5 s. The soft material can facilitate delicate grasping without damaging or deforming the structure, as seen in examples of a plant stem and flexible elastomer tubing shown in Figure 12A,F. The soft material of the arm can conform around the body of the object, making it easy to hold on to thicker objects such as a 3D-printed gear and plate in Figure 12B and 12D, respectively, all the way down to thin objects such as a thin ruler shown in Figure 12E. Finally, the use of multiple SMAs increases the weightlifting capability; Figure 12A–H show ascending order by weight of object held starting from 2 g up to 50 g. We can conclude that increasing the number of embedded muscles will further increase the weightlifting capabilities.

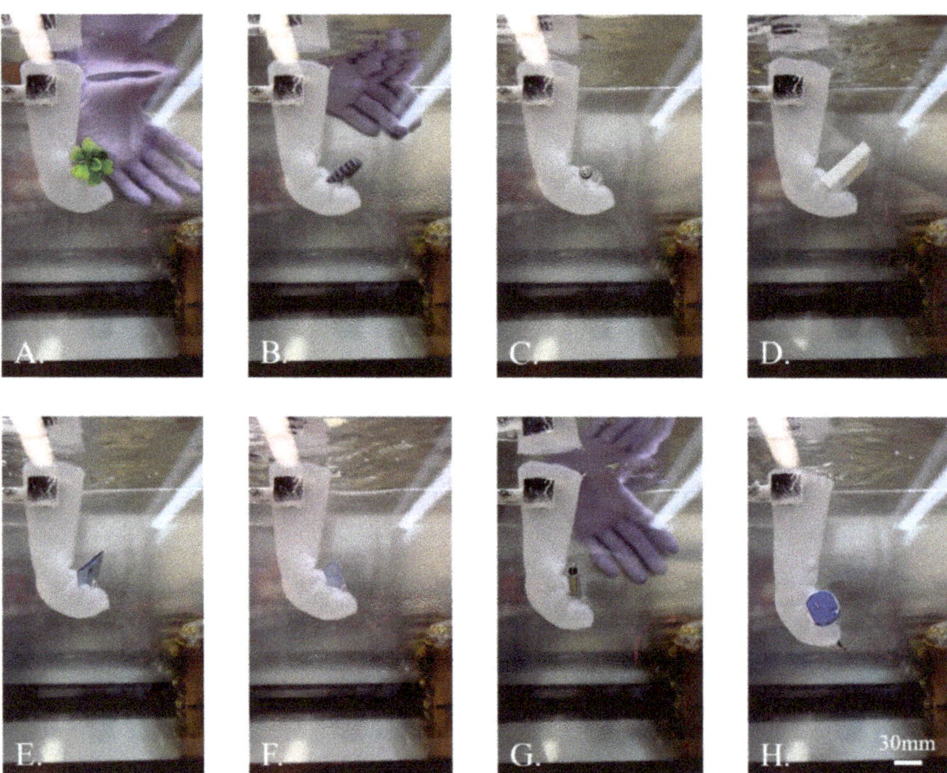

Figure 12. The grasping of objects. (**A**) a plant (2 g). (**B**) Small 3D-printed gear (10 g). (**C**) A plastic syringe (15 g). (**D**) 3D-printed flat plate (15 g). (**E**) Thin plastic ruler (15 g). (**F**) small plastic tube (20 g). (**G**) Metal spring (40 g). (**H**) A standard calibrated weight (50 g).

3.6. Shape Manipulation Utilizing Sequential Actuation

Shape manipulation was achieved by actuating SMAs in a series of combinations. We observed 2D and 3D bending, as shown in Figure 13. The first was utilizing all 3 embedded SMAs at the same time, the second was for 2 SMAs (long and short), the third was for the opposite 2 SMAs (long and short), and lastly for 2 SMAs (both short). Actuating 3 SMAs at the same time provided 3D bending where the arm initially starts bending in the XY plane and then transfers to the ZY plane. Actuating 2 SMAs (long and short) in the first case resulted in 2D bending in the XY plane, while the third case provided 2D bending in the ZY plane. We observe that, in this case, the bending plane is dependent on the muscle location but both will result in 2D manipulation. Actuating 2 SMAs (both short) results in 2D bending as well. Adding more SMAs will result in a more complex manipulation platform and higher weightlifting capabilities. An integrated control system can be developed to initialize the sequential actuation based on user input.

Figure 13. Snapshots of shape manipulation utilizing sequential actuation. (**A**) Activating 3 SMAs. (**B**) 2 SMAs (long middle and the one towards positive Z). (**C**) 2 SMAs (long middle and the one towards negative Z).

4. Conclusions and Future Work

In this paper, we have defined the design parameters of a biomimetic octopus-inspired arm actuated by coiled Shape Memory Alloys embedded inside Ecoflex 0010 silicone. Embedding the coiled Shape Memory Alloys gives rise to a curling motion for the whole arm. The strain in the coiled SMA is about 60%, giving rise to the arm achieving a curl throughout the length. The maximum bending angle that is achieved from the fabricated arm is 48°. The arm with one coiled SMA can achieve a maximum bending angle of about 17° when a load of 25 g is attached to the tip of the arm, about 14° when 50 g is attached, and about 10° when a load of 75 g is attached in an underwater environment. To achieve the curling motion and efficiently reach the bending angles mentioned above, a minimum of 140 J of electrical energy is required to be supplied to the arm. Future work is necessary for this robotic system to be deployed into the water for data acquisition. Topics such as continuum robotics, modeling, and simulation of the displacement of the tip of the arm can be implemented in the system. Some design changes to improve the performance of the arm are attainable. If multiple SMA coils with varying lengths are embedded in the silicone that forms the arm, a more controlled and hence better bending can be achieved, and the arm can be designed to have better grasping abilities and achieve more curl. Adding some temperature and vision sensors can improve the system further, which can then be used to grasp objects underwater delicately without causing much disturbance in the ecosystem.

Supplementary Materials: The following supporting information can be downloaded at https://www.mdpi.com/article/10.3390/act12100377/s1, Movie S1: Video showcasing actuation of the arm under described conditions.

Author Contributions: S.D. designed the soft arm, characterized the SMA, integrated the material, and tested the arm in a lab setting. S.D. and Y.A. developed the temperature predictive model. S.D. drafted the paper. Y.A. directed the research and contributed to the technical writing and editing of the manuscript. All authors have read and agreed to the published version of the manuscript.

Funding: This research was funded through an internal faculty startup research fund.

Data Availability Statement: Data are available through request.

Acknowledgments: We would like to thank the department of Mechanical Engineering at Wayne State University, Detroit, Michigan. Without the support of the department, this research work would not have been possible.

Conflicts of Interest: The authors declare no conflict of interest.

References

1. Tolley, M.T.; Shepherd, R.F.; Galloway, K.C.; Wood, R.J.; Whitesides, G.M. A resilient, untethered soft robot. *Soft Robot.* **2014**, *1*, 213–223. [CrossRef]
2. Shintake, J.; Shea, H.; Floreano, D. Biomimetic underwater robots based on dielectric elastomer actuators. In Proceedings of the 2016 IEEE/RSJ International Conference on Intelligent Robots and Systems (IROS), Daejeon, Republic of Korea, 9–14 October 2016; IEEE: New York, NY, USA, 2016; pp. 4957–4962.
3. Calderón, A.A.; Ugalde, J.C.; Zagal, J.C.; Pérez-Arancibia, N.O. Design, fabrication and control of a multi-material-multi-actuator soft robot inspired by burrowing worms. In Proceedings of the 2016 IEEE International Conference on Robotics and Biomimetics (ROBIO), Qingdao, China, 3–7 December 2016; IEEE: New York, NY, USA, 2016; pp. 31–38.
4. Kim, S.; Laschi, C.; Trimmer, B. Soft robotics: A bioinspired evolution in robotics. *Trends Biotechnol.* **2013**, *31*, 287–294. [CrossRef] [PubMed]
5. Almubarak, Y.; Punnoose, M.; Maly, N.X.; Hamidi, A.; Tadesse, Y. KryptoJelly: A jellyfish robot with confined, adjustable pre-stress, and easily replaceable shape memory alloy NiTi actuators. *Smart Mater. Struct.* **2020**, *29*, 075011. [CrossRef]
6. Almubarak, Y.; Schmutz, M.; Perez, M.; Shah, S.; Tadesse, Y. Kraken: A wirelessly controlled octopus-like hybrid robot utilizing stepper motors and fishing line artificial muscle for grasping underwater. *Int. J. Intell. Robot. Appl.* **2022**, *6*, 543–563. [CrossRef]
7. Almubarak, Y.; Tadesse, Y. Twisted and coiled polymer (TCP) muscles embedded in silicone elastomer for use in soft robot. *Int. J. Intell. Robot. Appl.* **2017**, *1*, 352–368. [CrossRef]
8. Li, J.; He, J.; Yu, K.; Woźniak, M.; Wei, W. A biomimetic flexible fishtail embedded with shape memory alloy wires. *IEEE Access* **2019**, *7*, 166906–166916. [CrossRef]
9. Fan, J.; Du, Q.; Dong, Z.; Zhao, J.; Xu, T. Design of the Jump Mechanism for a Biomimetic Robotic Frog. *Biomimetics* **2022**, *7*, 142. [CrossRef]
10. Potnuru, A.; Tadesse, Y. Characterization of coiled SMA actuators for humanoid robot. In *Active and Passive Smart Structures and Integrated Systems*; SPIE: Bellingham, WA, USA, 2017; pp. 175–184.
11. Hamidi, A.; Almubarak, Y.; Tadesse, Y. Multidirectional 3D-printed functionally graded modular joint actuated by TCPFL muscles for soft robots. *Bio-Des. Manuf.* **2019**, *2*, 256–268. [CrossRef]
12. Matharu, P.S.; Gong, P.; Guntaka, K.P.R.; Almubarak, Y.; Jin, Y.; Tadesse, Y.T. Jelly-Z: Swimming performance and analysis of twisted and coiled polymer (TCP) actuated jellyfish soft robot. *Sci. Rep.* **2023**, *13*, 11086. [CrossRef]
13. Song, S.-H.; Lee, J.-Y.; Rodrigue, H.; Choi, I.-S.; Kang, Y.J.; Ahn, S.-H. 35 Hz shape memory alloy actuator with bending-twisting mode. *Sci. Rep.* **2016**, *6*, 21118. [CrossRef]
14. Hamidi, A.; Almubarak, Y.; Rupawat, Y.M.; Warren, J.; Tadesse, Y. Poly-Saora robotic jellyfish: Swimming underwater by twisted and coiled polymer actuators. *Smart Mater. Struct.* **2020**, *29*, 045039. [CrossRef]
15. Haines, C.S.; Lima, M.D.; Li, N.; Spinks, G.M.; Foroughi, J.; Madden, J.D.; Kim, S.H.; Fang, S.; Jung de Andrade, M.; Göktepe, F. Artificial muscles from fishing line and sewing thread. *Science* **2014**, *343*, 868–872. [CrossRef] [PubMed]
16. Saharan, L.; de Andrade, M.J.; Saleem, W.; Baughman, R.H.; Tadesse, Y. iGrab: Hand orthosis powered by twisted and coiled polymer muscles. *Smart Mater. Struct.* **2017**, *26*, 105048. [CrossRef]
17. Matharu, P.S.; Ghadge, A.A.; Almubarak, Y.; Tadesse, Y. Jelly-Z: Twisted and coiled polymer muscle actuated jellyfish robot for environmental monitoring. *Acta IMEKO* **2022**, *11*, 1–7. [CrossRef]
18. Lee, J.-H.; Chung, Y.S.; Rodrigue, H. Long shape memory alloy tendon-based soft robotic actuators and implementation as a soft gripper. *Sci. Rep.* **2019**, *9*, 11251. [CrossRef]
19. Deng, E.; Tadesse, Y. A soft 3D-printed robotic hand actuated by coiled SMA. *Actuators* **2020**, *10*, 6. [CrossRef]
20. Calisti, M.; Giorelli, M.; Levy, G.; Mazzolai, B.; Hochner, B.; Laschi, C.; Dario, P. An octopus-bioinspired solution to movement and manipulation for soft robots. *Bioinspir. Biomim.* **2011**, *6*, 036002. [CrossRef]
21. Arienti, A.; Calisti, M.; Giorgio-Serchi, F.; Laschi, C. PoseiDRONE: Design of a soft-bodied ROV with crawling, swimming and manipulation ability. In Proceedings of the 2013 OCEANS, San Diego, CA, USA, 23–27 September 2013; IEEE: New York, NY, USA, 2013; pp. 1–7.

22. Cianchetti, M.; Calisti, M.; Margheri, L.; Kuba, M.; Laschi, C. Bioinspired locomotion and grasping in water: The soft eight-arm OCTOPUS robot. *Bioinspir. Biomim.* **2015**, *10*, 035003. [CrossRef]
23. Laschi, C.; Mazzolai, B.; Mattoli, V.; Cianchetti, M.; Dario, P. Design of a biomimetic robotic octopus arm. *Bioinspir. Biomim.* **2009**, *4*, 015006. [CrossRef]
24. Guo, Y.; Liu, L.; Liu, Y.; Leng, J. Review of dielectric elastomer actuators and their applications in soft robots. *Adv. Intell. Syst.* **2021**, *3*, 2000282. [CrossRef]
25. Marchese, A.D.; Onal, C.D.; Rus, D. Autonomous soft robotic fish capable of escape maneuvers using fluidic elastomer actuators. *Soft Robot.* **2014**, *1*, 75–87. [CrossRef] [PubMed]
26. Cheng, T.; Li, G.; Liang, Y.; Zhang, M.; Liu, B.; Wong, T.-W.; Forman, J.; Chen, M.; Wang, G.; Tao, Y. Untethered soft robotic jellyfish. *Smart Mater. Struct.* **2018**, *28*, 015019. [CrossRef]
27. Wehner, M.; Truby, R.L.; Fitzgerald, D.J.; Mosadegh, B.; Whitesides, G.M.; Lewis, J.A.; Wood, R.J. An integrated design and fabrication strategy for entirely soft, autonomous robots. *Nature* **2016**, *536*, 451–455. [CrossRef]
28. Joshi, A.; Kulkarni, A.; Tadesse, Y. FludoJelly: Experimental study on jellyfish-like soft robot enabled by soft pneumatic composite (SPC). *Robotics* **2019**, *8*, 56. [CrossRef]
29. Shepherd, R.F.; Ilievski, F.; Choi, W.; Morin, S.A.; Stokes, A.A.; Mazzeo, A.D.; Chen, X.; Wang, M.; Whitesides, G.M. Multigait soft robot. *Proc. Natl. Acad. Sci. USA* **2011**, *108*, 20400–20403. [CrossRef]
30. Frame, J.; Lopez, N.; Curet, O.; Engeberg, E.D. Thrust force characterization of free-swimming soft robotic jellyfish. *Bioinspir. Biomim.* **2018**, *13*, 064001. [CrossRef] [PubMed]
31. Gruber, D.F.; Wood, R.J. Advances and future outlooks in soft robotics for minimally invasive marine biology. *Sci. Robot.* **2022**, *7*, eabm6807. [CrossRef]
32. Sfakiotakis, M.; Kazakidi, A.; Tsakiris, D. Octopus-inspired multi-arm robotic swimming. *Bioinspir. Biomim.* **2015**, *10*, 035005. [CrossRef]
33. Wu, Q.; Yang, X.; Wu, Y.; Zhou, Z.; Wang, J.; Zhang, B.; Luo, Y.; Chepinskiy, S.A.; Zhilenkov, A.A. A novel underwater bipedal walking soft robot bio-inspired by the coconut octopus. *Bioinspir. Biomim.* **2021**, *16*, 046007. [CrossRef]

Article

An Optimized Design of the Soft Bellow Actuator Based on the Box–Behnken Response Surface Design

Jutamanee Auysakul [1,*], Apidet Booranawong [2], Nitipan Vittayaphadung [1] and Pruittikorn Smithmaitrie [1]

[1] Department of Mechanical and Mechatronics Engineering, Faculty of Engineering, Prince of Songkla University, Songkhla 90110, Thailand; nitipan.v@psu.ac.th (N.V.); pruittikorn.s@psu.ac.th (P.S.)
[2] Department of Electrical Engineering, Faculty of Engineering, Prince of Songkla University, Songkhla 90110, Thailand; apidet.b@psu.ac.th
* Correspondence: jutamanee.a@psu.ac.th

Abstract: Soft actuator technology is extensively utilized in robotic manipulation applications. However, several existing designs of soft actuators suffer from drawbacks such as a complex casting process, a multi-air chamber configuration, and insufficient grasping force. In this study, we propose a novel soft bellow design featuring a single air chamber, which simplifies the fabrication process of the actual model. To enhance the performance of the proposed design, we employ the Box–Behnken response surface design to generate a design matrix for implementing different levels of design factors in the finite element model. The FEA response is then subjected to an analysis of variance to identify significant factors and establish a regression model for deformation and stress response prediction. Among the considered responses, the wall thickness emerges as the most influential factor, followed by the divided ratio of radians and the number of bellows. Validation of the optimized soft bellow actuator's deformation response is performed through comparison with experimental data. Moreover, the soft bellow actuator is capable of exerting a pulling force of 8.16 N when used in conjunction with a simple gripper structure design, enabling effective object manipulation. Additionally, the soft bellow design boasts cost-effectiveness and easy moldability, facilitating seamless integration with different gripper frames for diverse applications. Its simplicity and versatility make it a promising choice for various robotic manipulation tasks.

Keywords: soft bellow actuator; soft gripper; FEA; response surface analysis; Box–Behnken design

Citation: Auysakul, J.; Booranawong, A.; Vittayaphadung, N.; Smithmaitrie, P. An Optimized Design of the Soft Bellow Actuator Based on the Box–Behnken Response Surface Design. *Actuators* **2023**, *12*, 300. https://doi.org/10.3390/act12070300

Academic Editor: Hamed Rahimi Nohooji

Received: 11 June 2023
Revised: 16 July 2023
Accepted: 22 July 2023
Published: 24 July 2023

1. Introduction

In the automation industry, robot manipulation plays a pivotal role in process lines. The end-effector of a robot arm is a crucial component responsible for manipulating target objects through grasping [1]. Traditional grippers are designed with a rigid structure to handle specific items. However, they are unsuitable for gripping delicate or deformable objects encountered in manufacturing, such as food and fruit products, which lack stable geometry and possess complex surface properties. To overcome this limitation, soft grippers have been developed, offering the advantage of securely grasping objects with uneven shapes and precise positioning without causing damage [2]. This type of actuator has significantly enhanced the interaction between robots and object surfaces, prioritizing safety. Its applications extend beyond industrial work [3–5] and encompass fields like rehabilitation [6–9] and harvesting [10–12].

Soft grippers utilize various types of actuators [13] to hold, handle, and grasp objects, including pneumatic, vacuum, shape memory alloy, and cable-driven actuators. Among these, pneumatic actuators are highlighted in this study due to their adaptability and rapid response [14]. Soft pneumatic actuators exhibit a range of designs suitable for gripper applications [15]. One common design is pneumatic networks (PenuNets) [16–19], which involve the creation of multi-air chambers that utilize the bending effect to grip objects.

However, the fabrication process of the real model using silicone parts can be complicated, requiring the gluing together of multiple parts. Additionally, this design tends to provide a lower lift force when holding objects. To address these challenges, a more easily fabricated design called the fiber-reinforced design [20–22] has been proposed. This design features a single-celled air chamber covered with silicone and winding fiber, utilizing a bending state to hold objects. Another design approach, known as granular jamming [23], employs a donut-shaped structure with two chambers controlled by air pressure and vacuum to enclose objects, enabling the gripping of various sizes and shapes. Gao et al. [24] introduced a telescopic design with passive retraction and a gripping force of 14 N. On the other hand, the pneumatic bellow actuator [25–27] relies on rapid deformation responses achieved through compression and friction forces to grip objects. This design allows for direct control of deformation by adjusting the air pressure and is simpler to cast as a single chamber component. However, it provides limited information regarding optimal factors influencing deformation and stress response. To address these limitations, this research proposes an optimized design for the soft bellow actuator.

Response surface methodology (RSM) is a commonly used design of experiments (DOE) approach to determine the optimal influential factors in a given problem [28,29]. By employing statistical and mathematical techniques along with analysis of variance (ANOVA), this method facilitates the optimization and prediction of responses based on a set of independent variables. The approximating function is typically divided into first- and second-order models. Several design methodologies, such as the $3K$ factorial, central composite, and Box–Behnken designs, are commonly employed for second-order models [30]. The main advantage of the Box–Behnken design is its efficiency in exploring the design space and determining the optimal factor settings with a relatively small number of experiments. Compared to full factorial designs, the Box–Behnken design requires fewer experimental runs while still providing accurate and reliable results. This is particularly beneficial when conducting experiments that involve time-consuming or expensive simulations or physical prototypes.

In the context of soft actuator applications, Cao et al. [26] introduced the Box–Behnken design to optimize the parameters of a double-chamber bellow actuator, validating the displacement response through experimental data. Honarpardaz M. [31] utilized the Latin Hypercube Sampling Design with a full quadratic model to optimize the design of PenuNets, focusing on minimizing stress while maximizing the gripper's payload. Chen et al. [32] utilized the topology optimization procedure to maximize the bending ability and enhance the payload capacity of the gripper. Finite element analysis (FEA) is a commonly employed technique to derive the response of an optimized design [33,34]. It serves as a valuable tool for enhancing the simulation model prior to manufacturing, saving time and eliminating the need for trial-and-error iterations.

This study aims to optimize the independent variables of the soft bellow actuator by employing a DOE approach with Box–Behnken response surface design. The objective is to achieve high deformation while minimizing stress. FEA is utilized to investigate the behavior of the soft bellow actuator in more than twenty cases. The proposed bellow design offers advantages in terms of fabrication process reduction by featuring a single air chamber and generating a significant pulling force that can be adapted for various applications. The framework presented in this work provides a systematic approach for researchers to optimize the soft actuator by reducing the design lead time and facilitating its application to other soft actuator models. The key objectives of this research can be summarized as follows:

- To study the parameters that influence the deformation response of the soft bellow actuator;
- To determine the optimized model of the soft bellow actuator and analyze the significant parameters using ANOVA;
- To propose a regression model to predict the deformation and stress response for this specific design of the soft bellow actuator;

- To validate the optimized design through experimental testing, assessing the deformation response and grasping performance using pulling force testing.

By addressing these objectives, this study provides valuable insights into the optimization of soft bellow actuators, enabling their efficient and effective use in various applications. The remainder of the article is organized as follows: Section 2 provides an overview of the workflow, the mathematical model of hyper-elastic materials, and the concept of DOE. Section 3 describes the geometric model of the soft bellow actuator and simulation analysis. Section 4 discusses significant parameters, the regression model, and the optimized design based on the Box–Behnken response surface design. Section 5 focuses on the manufacturing process and assembly of the soft bellow actuator. Section 6 presents the validation of the optimized design through deformation response and pulling-force performance. Finally, Section 7 concludes the paper, summarizing the findings and contributions of this study.

2. Materials and Methods

2.1. Workflow

The systematic workflow employed in this study is shown in Figure 1, which outlines the methodologies utilized. To assess the effectiveness of the optimized soft bellow actuator design, FEA is employed to identify the factors influencing deformation. These identified parameters are then applied to the Box–Behnken experimental design, generating a three-level design matrix. FEA is conducted using the MSC Marc Mentat Software (2020.0.0) to compute deformation and stress within the soft bellow actuator, allowing for the estimation of response parameters based on the generated design matrix. ANOVA is subsequently performed on the FEA results, calculating *p*-values to identify significant factors and establishing a regression model that correlates these factors with response parameters. By considering large deformation response and acceptable stress levels at a pneumatic pressure of 50 kPa, the optimal trial model within the design matrix is determined. To validate the deformation response of the optimized design, simulations are conducted and compared with experimental data. Subsequently, the soft bellow design is integrated into a rigid gripper structure to facilitate testing of its pulling force capabilities. By considering a large deformation response and acceptable stress levels at a pneumatic pressure of 50 kPa, the optimal trial model within the design matrix is determined. To validate the deformation response of the optimized design, simulations are conducted and compared with experimental data. Subsequently, the soft bellow design is integrated into a rigid gripper structure to facilitate testing of its pulling-force capabilities.

2.2. A Mathematical Model of Hyper-Elastic Materials

The soft bellow actuator incorporates a hyperelastic material for prototyping purposes, as it possesses the ability to undergo significant deformations when subjected to pressure, allowing for delicate object grasping [35]. In order to accurately characterize the hyperelastic behavior, nonlinear finite elements are employed. Various studies have proposed hyperelastic material models such as Mooney–Rivlin, Yeoh, Neo-Hookean, and Ogden by fitting experimental data curves. The numerical accuracy of these models relies on their order; however, a higher order does not necessarily guarantee a stronger correlation. For fundamental experiments like the uniaxial test, the two- or three-order Mooney–Rivlin approximation is commonly used [36]. The Mooney–Rivlin constitutive model is based on the strain energy function (W). In the absence of volumetric deformation and temperature variations, this function can be expressed in polynomial form using MSC Marc Mentat Software (2020.0.0) as

$$W = \sum_{i+j=1}^{n} C_{ij}(I_1 - 3)^i (I_2 - 3)^j \tag{1}$$

where I_1 and I_2 are the deformation invariants, whereas C_{ij} are material-specific constants. These deformation invariants can be expressed as follows:

$$I_1 = \lambda_1^2 + \lambda_2^2 + \lambda_3^2 \tag{2}$$

$$I_2 = \lambda_1^2\lambda_2^2 + \lambda_2^2\lambda_3^2 + \lambda_3^2\lambda_1^2 \tag{3}$$

where λ_1, λ_2, and λ_3 are the elongations of the element in three directions, respectively.

The hyper-elastic material can be assumed to be incompressible by:

$$\lambda_1^2\lambda_2^2\lambda_3^2 = 1 \tag{4}$$

The elongation in each direction is a function of strain (ε), which can be presented by

$$\lambda = 1 + \varepsilon \tag{5}$$

Thus, the deformation invariants for the uniaxial deformation mode can be expressed as follows:

$$I_1 = \lambda^2 + 2\lambda^{-1} \tag{6}$$

$$I_2 = 2\lambda + \lambda^{-2} \tag{7}$$

According to the Mooney–Rivlin three-order model, the stress (σ) of the hyperelastic material can be defined in terms of elongation and the C_{ij} parameter as follows:

$$\sigma(\lambda) = 2[C_{10}(\lambda - \lambda^{-2}) + C_{01}(1 - \lambda^{-3}) + 3C_{11}(\lambda^2 - \lambda - 1 + \lambda^{-2} + \lambda^{-3} - \lambda^{-4})] \tag{8}$$

It is important to note that the internal pneumatic pressure can influence the elongation behavior of a flexible bellow actuator. The level of air pressure plays a significant role in determining the deformation and stress response of the actuator.

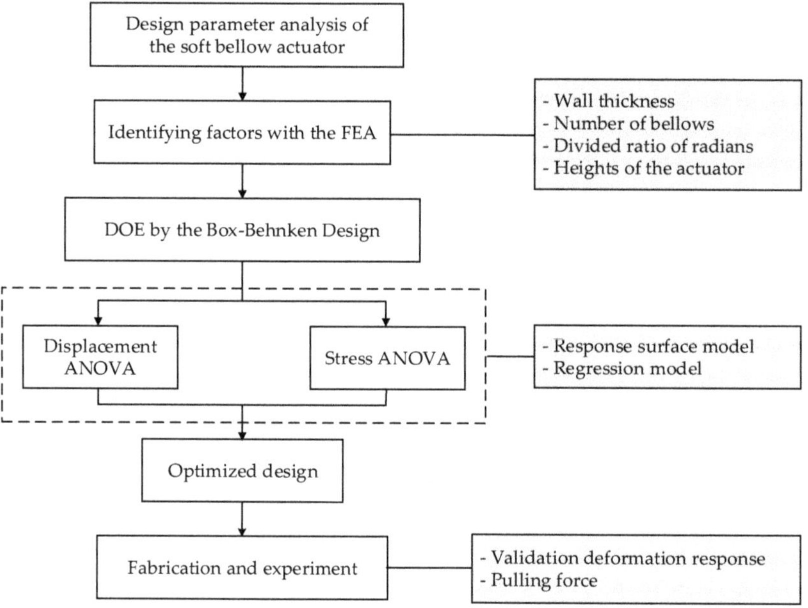

Figure 1. Analysis method.

2.3. Design and Statistical Analysis of Experiments

The Box–Behnken design is a well-known experimental design method that incorporates a response surface and is particularly suitable for investigating three independent variables in an experiment. Each factor is assigned code values representing the minimum, median, and maximum levels, resulting in three levels for each factor. These coded values facilitate the development of response surface designs. Additionally, to analyze the effect of parameters and estimate the regression equation, the Box–Behnken design response is employed. By considering the number of factors (K) and the number of center points (P), the total number of experiments (N) can be determined using the following equation:

$$N = K^2 + K + P \tag{9}$$

The Box–Behnken response surface is generalized to the second order of the polynomial regression mode [30] as follows:

$$Y = \beta_0 + \sum_{i=1}^{k} \beta_i x_i + \sum_{i=1}^{k} \beta_{ii} x_i^2 + \sum_{i=1}^{k-1} \sum_{j=i+1}^{k} \beta_{ij} x_i x_j + \gamma \tag{10}$$

where Y is the response parameter; k is the number of independence variables (k = 3); x_i and x_j are the independence parameters; β_0 is the intercept coefficient; and β_i, β_{ii}, and β_{ij} are the regression coefficients for linear, quadratic, and interaction variables, respectively. The experimental miscalculation (γ) is assumed to be zero. This regression model was used to estimate the factor responses. The fitness of the polynomial regression is defined by the coefficient of determination (R^2). The significance of the factors is evaluated using ANOVA, employing the f-value at a specified p-value threshold. The optimal conditions for each factor are estimated through three-dimensional (3D) response surface plots and contour plots. Subsequently, the optimized model of the soft bellow actuator was validated by conducting experiments to verify the response.

3. Finite Element Analysis of Soft Bellow Design

3.1. Geometric Model of the Soft Bellow Actuator

The proposed design of the bellow actuator is shown in Figure 2. In this design, certain dimensions remain constant: the diameter on the top surface (D_1) is 40 mm, the height of the soft actuator (H_1) is 35 mm, and the length from the fixed base (H_2) is 5 mm. The height of the bellows layer (h) is determined by the number of bellows (N) and the wall thickness (t). The relationship between these parameters can be expressed as follows:

$$h = \frac{H_1 - H_2 - t}{N} \tag{11}$$

where h is used to compute the bellows' radian (r) using the following formula:

$$r = \frac{h}{d} \tag{12}$$

where d is the divided ratio of h that must be more than half of h in order to calculate the r of the soft bellow actuator. In order to facilitate general usage for object grasping and to simplify the optimization process, the contact surface of the model is designed to be flat.

3.2. Simulation Analysis

FEA is a widely used tool for predicting the behavior of a system, including deformation, stress, and more. In this study, MSC Marc Mentat Software (2020.0.0) is utilized to perform FEA simulations. To reduce computational costs, an axisymmetric approach is employed to calculate the effects of deformation and stress. The model utilizes quadrilateral mesh elements with four nodes, as depicted in Figure 3.

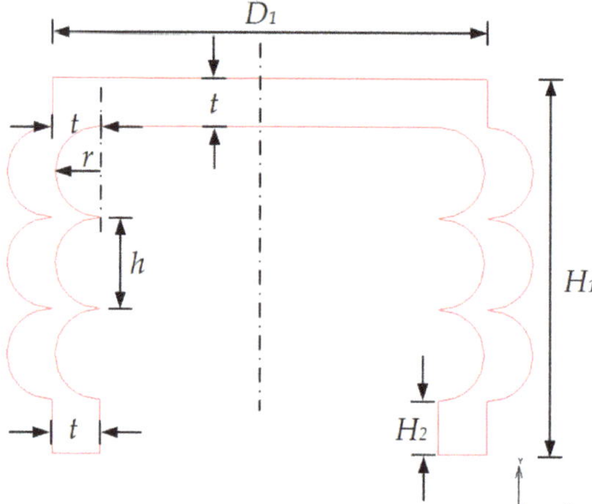

Figure 2. The geometry of the bellow design.

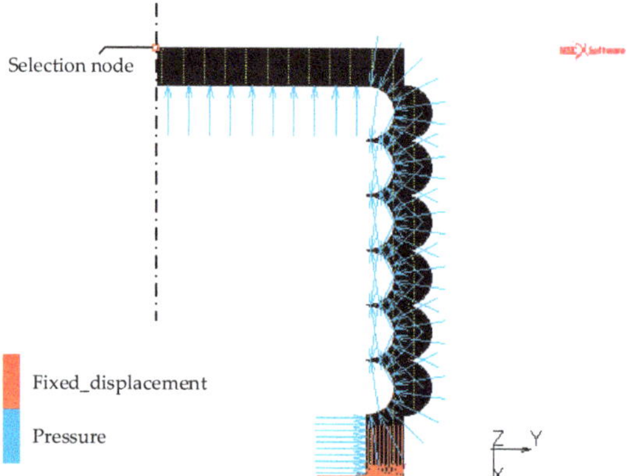

Figure 3. Boundary condition of an actuator.

Figure 3 shows the boundary conditions of the actuator with an axisymmetric model. In the analyzed model, the element type employs the full integration (Quad 10) formulation for solid structures. Given that the soft gripper deforms the actuator through pressurized air, several assumptions are made: (a) both the air pressure and atmospheric pressure are constant; (b) the actuator is completely sealed without any leaks; and (c) the pressure within the bellow is uniformly distributed and remains constant at each point. These assumptions allow for the definition of boundary conditions in the model. The ground of the actuator is subjected to a fixed displacement, while the inner surface of the curve generates pressure through an edge load, as illustrated in Figure 3. In this study, the edge load is set at 50 kPa, and a linear ramp table is employed to calculate the large strain of the static model with a constant time of 0.02, utilizing 50 steps. For model optimization, the deformation and stress response at the selected node, as shown in Figure 3, are considered. This particular location is crucial as it represents the weakest area responsible for supporting the object during grasping applications.

When it comes to applications involving soft actuators, materials like Ecoflex, Dragon Skin, and TPE rubber are commonly used for casting models. In this study, Dragon Skin 30 (Smooth-On, Inc., Macungie, PA, USA) was chosen as the material to construct the bellow actuator due to its desirable characteristics, including high tensile strength and significant elongation deformation, which make it well-suited for gripper applications. The material properties of Dragon Skin 30 were evaluated for tensile strength using the ASTM D412 standard, employing a Z005 tensile testing machine (Zwick Roell, Ulm, Germany). Five silicone samples of Die C, each with a thickness of 3 mm, were subjected to extension at a rate of 500 mm/min until rupture. Figure 4 presents a comparison between the experimental data and the fitted data for the stress and strain of Dragon Skin 30. The maximum tensile strength was determined to be 3.64 ± 0.21 MPa, while the elongation at break was measured to be 473 ± 28%.

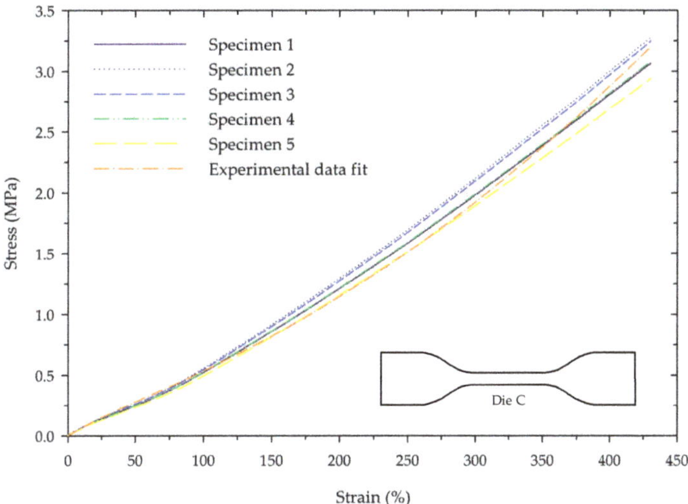

Figure 4. The tensile strength test data and experiment data fit.

To determine the material properties, data from specimen 1 was selected for its average characteristics compared to the other specimens. The material properties were computed using MSC Marc Mentat software (2020.0.0). The Mooney–Rivlin model of the third order was utilized, with the following coefficients: C_{10} = 1.192 × 10^{-1} MPa, C_{01} = 3.464 × 10^{-9} MPa, and C_{11} = 1.484 × 10^{-2} MPa, with an error 3.370 × 10^{-4}. To optimize the design of the soft bellow gripper, the deformation and stress at the center of the model's upper surface were analyzed. These response parameters were crucial for the DOE using the Box–Behnken method, allowing for the design of the bellow soft gripper with acceptable levels of deformation and stress.

In this study, we consider the following factors to influence the deformation of the soft bellow actuator: wall thickness, number of bellows, bellow radian, and actuator height. These factors are crucial in defining the response prior to applying the Box–Behnken method for factor reduction. The wall thickness is examined at three levels: 3 mm, 4.5 mm, and 6 mm. The casting process imposes a minimum wall thickness requirement, as thin walls are prone to tearing.

The simulation models in our study maintain a consistent number of bellows (six) and a divided ratio for the bellow radian (three). Figure 5 demonstrates the finite element analysis result of the comparative model. The lower part of the model, represented by the blue color, indicates fixed displacement, while the middle portion of the model's upper surface, depicted in red, exhibits significant displacement. This region is influenced by the internal pressure of the actuator and is located away from the fixing area.

Figure 6a presents the relationship between deformation displacement and pressure for comparable wall thicknesses. At a pressure of 50 kPa, the 3 mm wall thickness demonstrates the highest deformation displacement of 14.58 mm, whereas the 6 mm wall thickness exhibits the smallest deformation displacement of 5.49 mm. It is evident that the thin wall thickness significantly affects the deformation of the soft actuator. However, it is important to note that excessive stress on a thin wall may lead to actuator rupture.

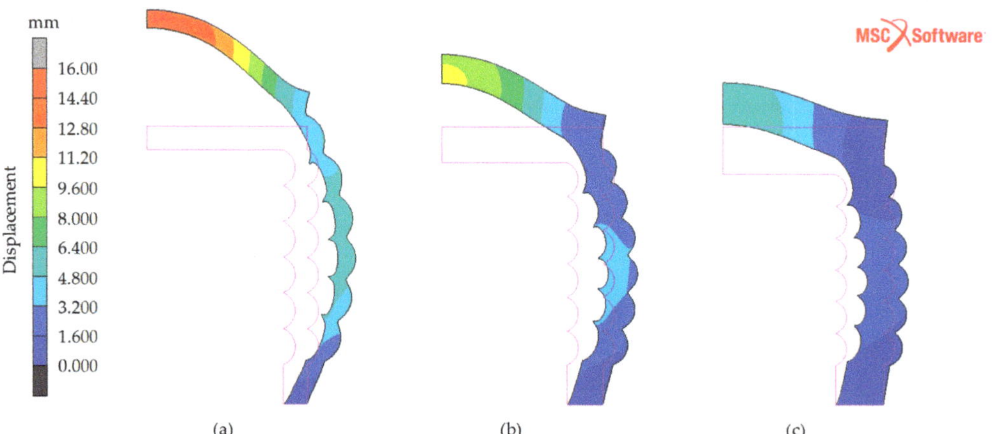

Figure 5. Simulation result of deformation response under 50 kPa pressure with different wall thicknesses of (**a**) 3 mm, (**b**) 4.5 mm, and (**c**) 6 mm.

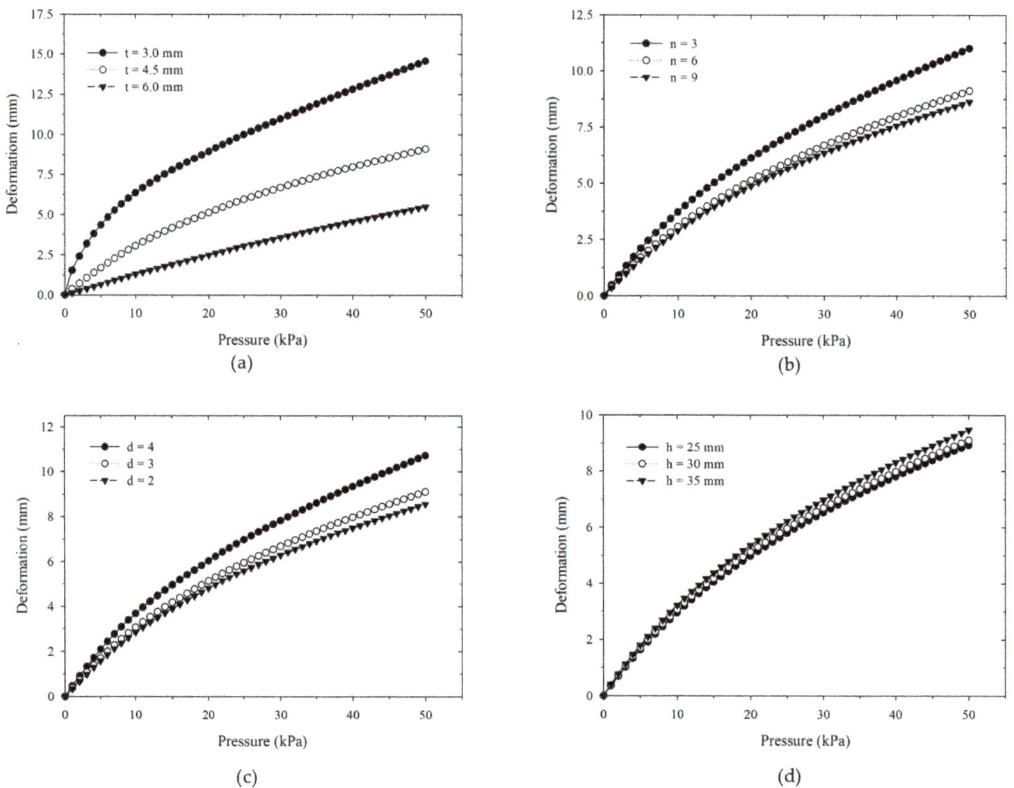

Figure 6. Relationship between deformation and air pressure of simulation results (**a**) under different wall thicknesses, (**b**) under different numbers of bellows, (**c**) under different divided ratios, and (**d**) under different heights of the actuator.

We compare the deformation response for three different numbers of bellows: three, six, and nine, simulated under the same conditions, as previously described. The models for comparison have the same wall thickness and a divided ratio of 4.5 mm and 3, respectively. In Figure 6b, it can be observed that the deformation with varying numbers of bellows at six and nine under the same pressure is not dissimilar. When the applied pressure is 50 kPa, the greatest deformation displacement of 11.0 mm is observed with three bellows. The deformation responses of these three cases are comparable, as the rigidity of the simulation model does not vary significantly.

Next, we examine the effect of the bellow's radian with different divided ratios of two, three, and four while keeping the same wall thickness (4.5 mm) and number of bellows (six). Figure 6c demonstrates the influence of the bellow's radian on the deformation response. The divided ratio of four yields the greatest deformation of 10.72 mm compared to the divided ratios of two and three. However, most of the models exhibit deformations greater than 8 mm, which is beneficial when adapting the model to a soft actuator for a wide range of object grasping.

Furthermore, when comparing the height of the actuator at 25 mm, 30 mm, and 35 mm with the same wall thickness (4.5 mm), number of bellows (six), and divided ratio of radians (three), Figure 6d illustrates that the actuator's height has no significant effect on the deformation response. The nearly identical deformations observed in each air-pressurized simulation case confirm that the actuator's height does not significantly influence the deformation response. Hence, the factors that have a notable impact on the deformation of the actuator include the wall thickness, number of bellows, and divided ratio of radians. These variables were used to create an experiment using the Box–Behnken method, allowing for the optimization of the model.

4. Box–Behnken Design

The optimization of the soft bellow actuator's design focuses on the deformation and stress at the contact surface. To achieve this goal, the Box–Behnken technique is employed to optimize the model's design through a matrix design. This method utilizes a response surface to determine the effect of factors and levels on the objective function. In this study, the deformation of the soft gripper was influenced by the wall thickness, number of bellows, and divided ratio of radians, as described in Section 2.3. In the Box–Behnken design, these factors are used to evaluate the surface response, with the index A representing the wall thickness (mm), B representing the number of bellows, and C representing the divided ratio of radians. The deformation and stress results at the selection node are denoted by D (mm) and E (kPa), respectively. Table 1 provides an overview of how the factors are classified into three levels.

Table 1. Design factors of the Box–Behnken method.

Factor	Name	Low	Median	High
A	Wall thickness	3	4.5	6
B	Number of bellows	3	6	9
C	Divided ratio of radians	2	3	4

The response surface design of the soft bellows is investigated using three factors and three levels. Equation (9) is used to calculate the total number of experiments, and the replicate number is three. As a result, the simulation test employs the design factors based on the Box–Behnken design, as shown in Table 2. The number of trials is equal to the number of experiments, totaling 15 cases. This parameter is utilized to define an optimum soft bellow design with high deformation and minimal stress.

Table 2. Box–Behnken design and response results.

Standard Order	Running Order	Point Type	Blocks	*A*	*B*	*C*	*D*	*E*
1	1	2	1	3	3	3	18.06	290.8
2	2	2	1	6	3	3	6.49	243.9
3	3	2	1	3	9	3	13.28	287.9
4	4	2	1	6	9	3	5.19	234.1
5	5	2	1	3	6	2	19.02	288.2
6	6	2	1	6	6	2	6.37	250.1
7	7	2	1	3	6	4	13.19	289.7
8	8	2	1	6	6	4	5.14	232.0
9	9	2	1	4.5	3	2	16.07	294.6
10	10	2	1	4.5	9	2	9.52	294.3
11	11	2	1	4.5	3	4	9.50	288.6
12	12	2	1	4.5	9	4	8.27	290.8
13	13	0	1	4.5	6	3	9.12	291.8
14	14	0	1	4.5	6	3	9.12	291.8
15	15	0	1	4.5	6	3	9.12	291.8

Analysis of Variance

The response surface with the Box–Behnken design is utilized to determine the influence of the deformation and stress of the bellow actuator, as observed from the FEA results presented in Table 2. The displacement response is subjected to ANOVA at a 95% confidence level, and the results are summarized in Table 3. The model *f*-value of 273.89 and the associated *p*-value ($p < 0.0001$) indicate that the regression model is statistically significant. The mean square (MS) value for the uncontrolled aspect of the simulation data had an inaccuracy of 0.1127. The analysis of the model's significant factors suggests that the *p*-values should be less than 0.05, with variables having lower *p*-values being more significant.

Table 3. Results of the ANOVA for extraction displacement of the soft bellow actuator.

Source	DF	Adj SS	Adj MS	F-Value	*p*-Value
Model	9	277.84	30.87	273.89	<0.0001
Linear					
A	1	203.59	203.59	1806.27	<0.0001
B	1	23.97	23.97	212.64	<0.0001
C	1	27.68	27.68	245.6	<0.0001
Square					
A×	1	2.76	2.76	24.49	0.0043
B×B	1	2.22	2.22	19.66	0.0068
C×C	1	3.31	3.31	29.38	0.0029
2-way Interaction					
A×B	1	3.05	3.05	27.02	0.0035
A×C	1	5.3	5.3	47.01	0.001
B×C	1	7.06	7.06	62.65	0.0005
Error	5	0.5636	0.1127		
Total	14	278.4			
R^2 = 0.998, adj-R^2 = 0.9943, C.V.% = 3.2, and Adeq Precision = 51.55					

As seen in Table 3, all variables, including linear, quadratic, and two-way interactions, have *p*-values less than 0.05, indicating their significant influence on the displacement response. The Adeq Precision is a measure of signal-to-noise ratio precision, with values greater than four indicating high precision. In this case, the Adeq Precision is 51.55, suggesting that the design space navigation model had a high level of precision. Through the analysis of displacement simulation data using multiple regression, the response model in the second-order polynomial equation can be predicted using Equation (13):

$$D = 76.26 - 10.287A - 3.811B - 13.65C + 0.3843A \times A + 0.0861B \times B + 0.947C \times C + 0.1939A \times B + 0.767A \times C + 0.4429B \times C \quad (13)$$

The R^2 and adj-R^2 values for this regression model are 0.998 and 0.9943, respectively. These values exceed 0.7, indicating a significant correlation between the displacement response and the independent variables. R^2 values closer to 1 suggest that the regression model is well-suited to the actual response.

Figure 7 presents the 3D response surface and contour plot of the regression equation (Equation (13)). These plots provide insights into the relationship between the displacement response and the levels of two independent variables. The results of the extraction affected by the wall thickness, number of bellows, and divided ratio of radians are displayed in Figure 7a–c. Using two continuous variables and fixing the other variable at its middle, Figure 7a illustrates the displacement response as a function of wall thickness and number of bellows while keeping the radian ratio constant at three. These two factors had a significant impact on increasing the deformation of the bellow actuator. The response range increases rapidly from 9.12 to 18.06 mm when the wall thickness is reduced below 4.5 mm. In Figure 7b, the effect of the divided ratio of radians and the number of bellows is shown, with the wall thickness kept constant at 4.5 mm. The displacement response increases as the parameter values decrease.

However, when the divided ratio of radians exceeds four, the displacement response of the model is restricted to a range of 13 mm. Figure 7c presents the response of the divided ratio of radians and the wall thickness, with the number of bellows kept constant at six. Reducing the values of these two factors leads to an increase in the deformation response within the range of 8.58 to 15.76 mm. Therefore, the regression model response highlights the significant influence of wall thickness on the displacement response, as the stiffness of the actuator depends on this factor. It is followed by the divided ratio of radians and the number of bellows.

Table 4 presents the ANOVA results based on the experimental stress data of the bellow actuator using the Box–Behnken design. The regression model exhibits a p-value close to zero and a high f-value of 119.83, indicating its overall effectiveness and significance. Factors A, C, $A \times A$, and $A \times C$ have high significance for the stress response, as their p-values are less than 0.05, while other factors have negligible effects. The regression model for the stress response of the soft bellow actuator is expressed by Equation (14).

$$E = 85.2 + 105.06A + 0.94B + 6.92C - 12.146A \times A - 0.027B \times B + 0.55C \times C - 0.381A \times B - 3.280A \times C + 0.215B \times C \quad (14)$$

The regression model for stress prediction demonstrates a good fit to the experimental data, with R^2 and adj-R^2 values of 0.9954 and 0.9871, respectively. This indicates that the model accurately predicts the stress of the actuator.

Figure 8 shows the response of the deformation and stress of the soft bellow actuator based on the trial numbers in the Box–Behnken design. The models with thick walls of 6 mm (trial numbers 2, 4, 6, and 8) exhibit a deformation response of less than 6.5 mm and a minimum stress in the range of 230–250 kPa. These models have high wall stiffness, resulting in lower deformation and stress. For the model with a wall thickness of 4.5 mm, trial numbers 10 to 15 show a deformation response of less than 10 mm, except for trial number 9. Trial number 9 has the minimum values for the number of bellows and the divided ratio of radians, leading to a higher deformation response. However, the stress response of all models with a 4.5 mm wall thickness is approximately 290 kPa.

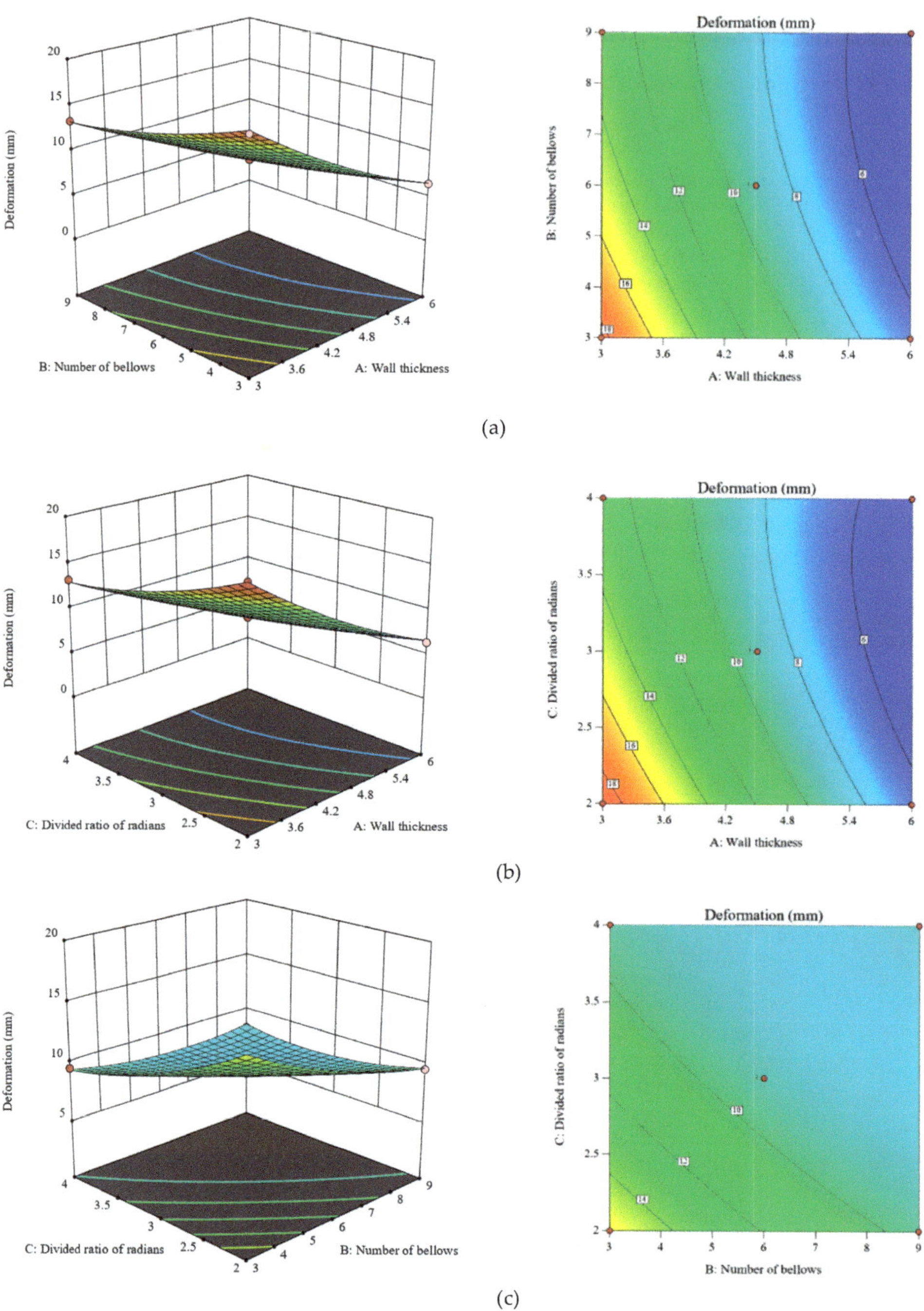

Figure 7. Response surface and contour diagrams of the soft bellow: (**a**) under constant the divided ratio of radians, (**b**) under constant the number of bellows, and (**c**) under constant the wall thickness.

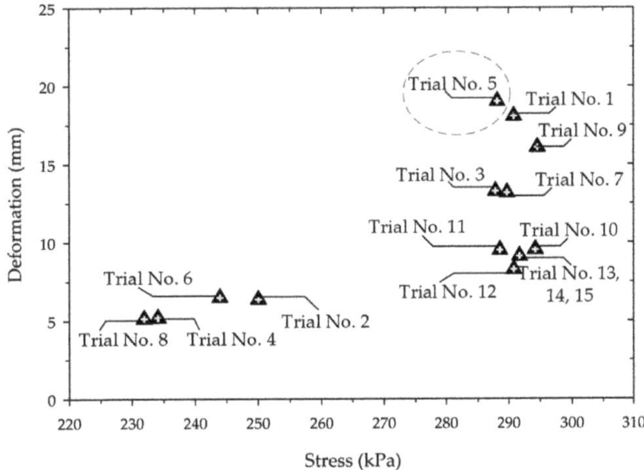

Figure 8. The Box–Behnken design matrix responds to stress and deformation.

Table 4. Results of the ANOVA for the extraction stress of the soft bellow actuator.

Source	DF	Adj SS	Adj MS	F-Value	*p*-Value
Model	9	7830.61	870.07	119.83	<0.0001
Linear					
A	1	4827.23	4827.23	664.82	<0.0001
B	1	14.86	14.86	2.05	0.212
C	1	84.18	84.18	11.59	0.0191
Square					
A×A	1	2757.63	2757.63	379.79	<0.0001
B×B	1	0.2251	0.2251	0.031	0.8671
C×C	1	1.12	1.12	0.1541	0.7108
2-way Interaction					
A×B	1	11.73	11.73	1.62	0.2597
A×C	1	96.82	96.82	13.33	0.0147
B×C	1	1.67	1.67	0.2299	0.6518
Error	5	36.3	7.26		
Total	14	7866.92			
R^2 = 0.9954 and adj-R^2 = 0.9871, C.V.% = 0.9716, and Adeq Precision = 29.572					

In contrast, the models with a thin wall thickness of 3 mm (trial numbers 1, 3, 5, and 7) exhibit a deformation response greater than 13 mm. Among these models, trial number 5, with the number of bellows and the ratio of radians is six and two, respectively. It demonstrates a significant amount of deformation. However, trial number 5 has a lower stress response compared to trial number 1 and 9, due to its greater rigidity. Therefore, trial number 5 represents the optimal design for the actuator, as it achieves high displacement while maintaining an appropriate stress level.

5. Manufacturing and Assembly

In this section, we proceeded to construct the physical model based on the optimized design from trial number 5 in the previous section. The optimized model has a wall thickness of 3 mm, six bellows, and a divided ratio of radians of two. The fabrication process of the soft bellow gripper involves two main components: the soft bellow part and the rigid frame structure. For the soft bellow part, a molding process is employed using the hyper-elastic material. This design features a single air chamber, which simplifies the fabrication process. The trial number 5 model has a volume of 16,567 mm^3 and weighs 17.89 g, considering the density of Dragon Skin 30 at 1.08 g/cm^3. Figure 9 shows the four steps involved in the casting process. The model's mold is divided into five pieces for easy disassembly and was created using a 3D printer, specifically the Flashforge Adventurer 3 (Zhejiang Flashforge 3D Technology Co., Ltd., Jinhua, China).

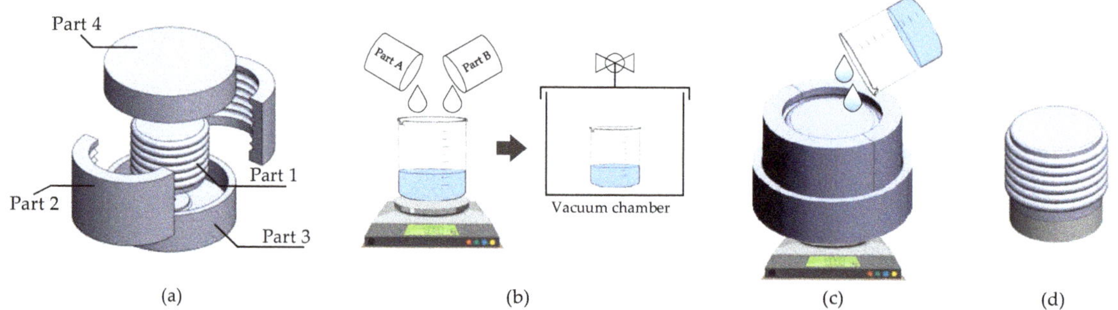

Figure 9. Fabrication of a Soft Bellow Actuator: (**a**) 3D-printed molds, (**b**) silicone preparation, (**c**) casting process, and (**d**) final assembly.

In the assembly steps, Part 1 of the mold is attached to Part 3 using a screw. To ensure proper fitting during mold construction, a tolerance of 10% is maintained between Part 2 and Part 3. Next, Dragon Skin 30, consisting of Part A and Part B in a 1:1 weight-to-weight ratio, is prepared. The silicone compound is thoroughly mixed, and any bubbles in the mixture are eliminated using a vacuum chamber. The liquid silicone is carefully poured into the mold, with the mass of the silicone being measured accurately. An additional 5% weight is added to account for any potential mold leakage. To ensure the optimal quality of the final product, the mold is shaken to eliminate any trapped air bubbles within the silicone. Once the pouring and shaking process is complete, the top surface of the mold is securely closed using Part 4. Subsequently, the silicone is left to cure for approximately 16 h at room temperature. Alternatively, it can be cured at a higher temperature to expedite the curing process.

The rigid frame of the gripper is specifically designed to enable the grasping and holding of objects, resembling the functionality of human fingers. For the purpose of this study, only two soft bellow actuators were utilized to evaluate the basic grasping properties of the optimized actuator. The frame structure, as depicted in Figure 10, consists of three individual components: the support plate for the soft bellow component, the gripping frame, and the holding plate. All of these parts were fabricated using 3D printing technology. The support plate serves the purpose of securely attaching the soft bellow component to the rigid part. Additionally, it includes a hose connector that supplies pressurized air to the soft actuator. To ensure a strong bond between the silicone part and the rigid part, an ethyl-based instant glue called Loctite 495, manufactured by Henkel AG & Co. KGaA, in Dusseldorf, Germany, is applied. The gripping frame is specifically designed to facilitate the grasping of objects with two fingers, accommodating objects ranging from 30 to 50 mm in size. It is connected to the holding plate, which serves the dual function of providing support for the end-effectors of the robot arm (Dobot CR5) during experimental procedures and enabling future applications such as pick-and-place operations.

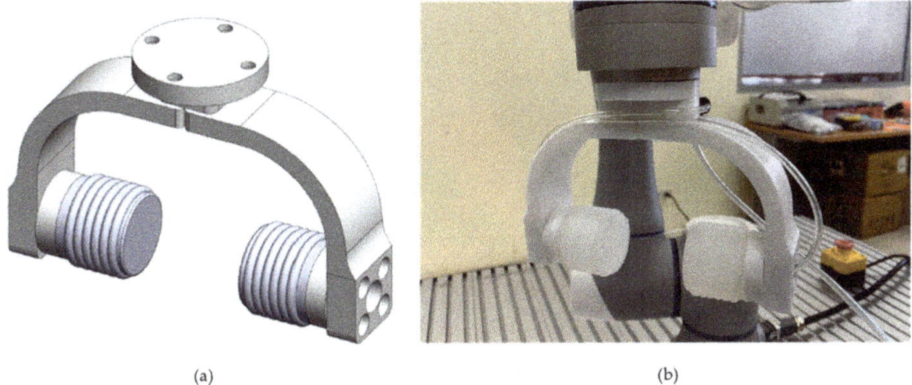

Figure 10. Assembly of the rigid frame and the soft bellow actuator (**a**) in the 3D model and (**b**) in the real model.

6. Experiments and Results

This study focused on evaluating the performance of the soft bellow actuators and the associated pulling force. To conduct these tests, the air pressure was carefully regulated using a digital regulator, specifically the SMC ITV2050 model. This regulator offers a pressure range of 0.005 to 0.9 MPa and has a quick response time of 0.1 s. To control the actuated air pressure, a voltage divider circuit was employed. This circuit allows for the adjustment of the output voltage within the range of 0 to 10 VDC using a potentiometer. By varying the output voltage, the desired air pressure for actuation can be achieved. The electrical and pneumatic systems are interconnected, as illustrated in Figure 11, providing a schematic representation of their setup and integration.

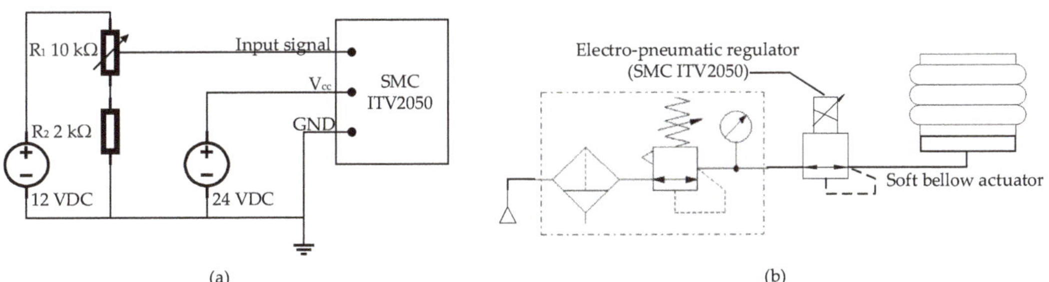

Figure 11. Schematic diagram of (**a**) the electrical system and (**b**) the pneumatic system.

6.1. Validation of the Deformation Response

The response of the soft bellow actuator to fluctuations in internal air pressure was evaluated to verify the accuracy of the optimized simulation model. The pneumatic pressure was adjusted in 5 kPa increments from 0 to 50 kPa. The displacement of the actuator was measured at steady internal air pressure conditions. This experiment was conducted five times with five samples to investigate the deformation response of the actuator. Figure 12 illustrates the deformation performance of the optimized soft bellow actuator at different internal air pressures. The actuator exhibited inflation as the pressure increased, with a maximum pressure limit of 50 kPa to prevent damage to the bellows.

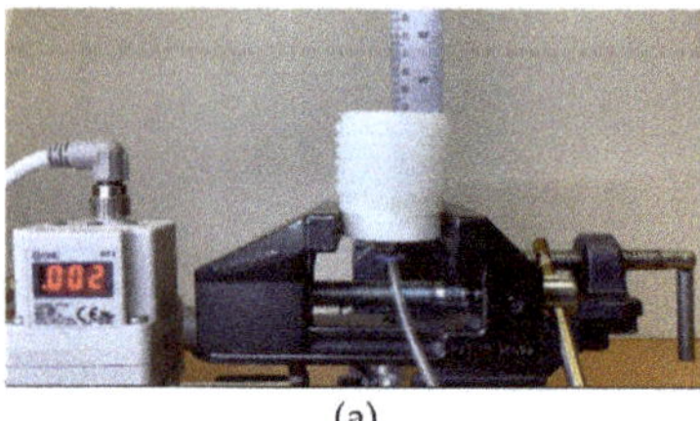

(a)

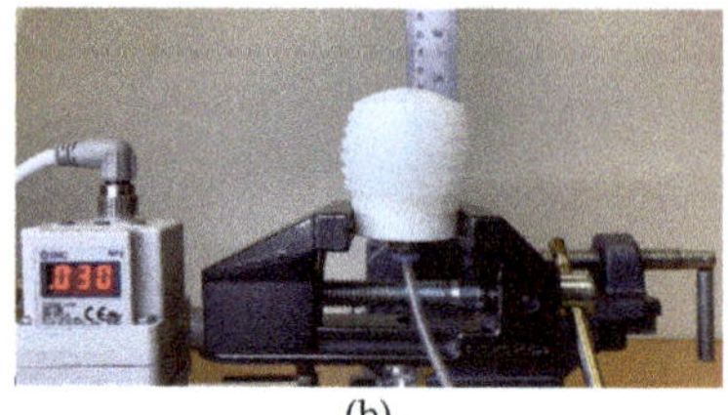

(b)

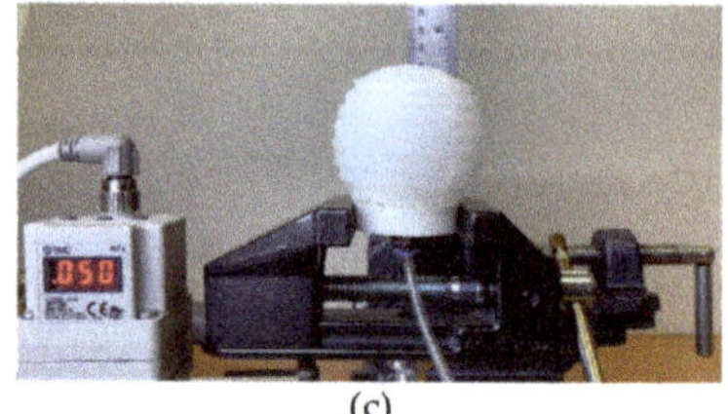

(c)

Figure 12. The deformation response at different air pressure levels: (**a**) at 2 kPa, (**b**) at 30 kPa, and (**c**) at 50 kPa.

Figure 13 shows a comparison of the deformation responses between the experimental and simulation results. In the pneumatic pressure range of 0 to 20 kPa, the experimental deformation showed a high error compared to the simulation, with an average error of 43.12%, or 3.22 mm. However, for pressures exceeding 25 kPa, the deformation of both the experiment and simulation results became more similar, with an error range of 2.51–2.18 mm or an average error of 14.54%. These discrepancies can be attributed to several factors, including material properties, manufacturing tolerances, boundary conditions, and simplified assumptions.

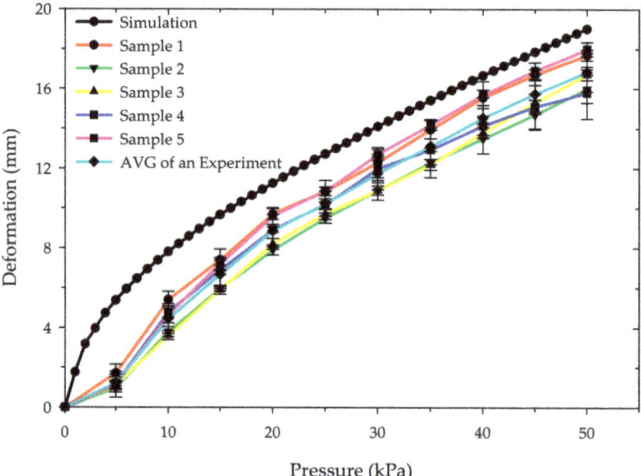

Figure 13. Comparison of deformation responses.

The material properties in the simulation, such as the elastic modulus and strain energy function of the soft bellow actuator of the Mooney–Rivlin model, may not accurately represent the actual properties, leading to differences between the predicted and experimental deformations. Additionally, the simulation assumes an idealized model without considering manufacturing imperfections, such as uneven thickness or slight variations in material properties, which can impact the behavior of the actuator and result in deviations from the experimental results. The boundary conditions in the simulation, which aim to replicate the experimental setup, may not precisely mimic the actual conditions, further contributing to differences in deformation. The simplified assumptions made in the simulation, particularly using an axisymmetric model, can also introduce inaccuracies, especially in scenarios involving intricate or nonlinear deformations at low pressures.

Furthermore, Table 5 provides a comparison of the deformation and strain between the experimental and simulation results. The experimental deformation reached 16.84 mm, while the analytical model was limited to 19.02 mm, resulting in a difference of less than

2.18 mm compared to the simulation. The strain values were 0.481% and 0.543% for the experiment and simulation, respectively, with an error of 11.46%. These variations can be attributed to the aforementioned factors influencing the inflation behavior of the soft bellow actuator, leading to the experimental response being lower than the simulation result. However, both the experimental and simulation results demonstrate a satisfactory deformation response.

Table 5. The deformation and strain of the optimized design.

Pressure (kPa)	Deformation (mm)		Strain (%)		% Error of a Deformation
	Experiment (Average)	Simulation	Experiment (Average)	Simulation	
5	1.22	5.37	0.035	0.153	−77.29
10	4.452	7.81	0.127	0.223	−42.99
15	6.672	9.66	0.191	0.276	−30.92
20	8.864	11.26	0.253	0.322	−21.29
25	10.224	12.73	0.292	0.364	−19.68
30	11.76	14.11	0.336	0.403	−16.63
35	13.124	15.41	0.375	0.440	−14.85
40	14.544	16.66	0.416	0.476	−12.72
45	15.74	17.86	0.450	0.510	−11.89
50	16.84	19.02	0.481	0.543	−11.46

6.2. Pulling force Experiment

The experimental setup for measuring the pulling force of the soft bellow gripper is depicted in Figure 14. The gripper, designed as per the proposed model, is mounted on a robot arm. Various objects, including a cube, a cylinder, and a sphere with a diameter of 50 mm, were grasped by the soft bellow gripper. To measure the pulling force, a digital force gauge (SF-100 Model, Wenzhou Sanhe Instrument Co., Ltd., Wenzhou, China) was attached to the tested object. The response of the gripper was evaluated at different levels of internal air pressure, ranging from 20 to 40 kPa.

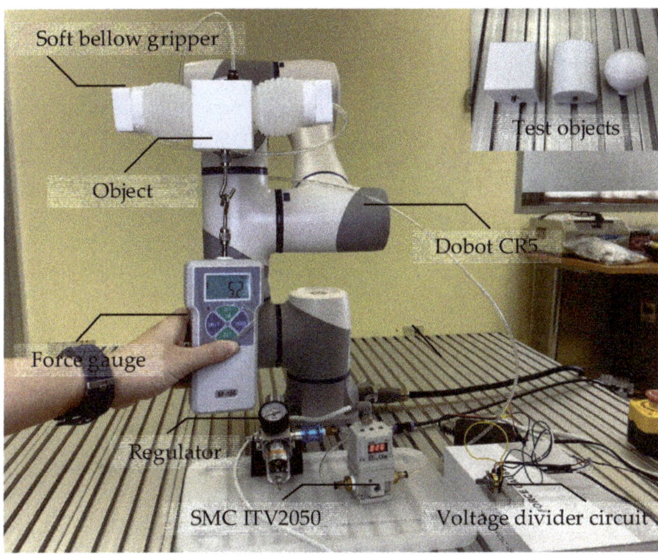

Figure 14. Experimental setup for pulling with different objects.

The response of the gripper was evaluated at different levels of internal air pressure, ranging from 20 to 40 kPa. Figure 15 presents the performance of the gripper in holding objects at different pressure levels. It can be observed that as the air pressure increased, the deformation of the actuator applied a compressive force to the object. However, excessive pneumatic pressure exceeding 40 kPa can adversely affect the contact force, potentially leading to actuator damage.

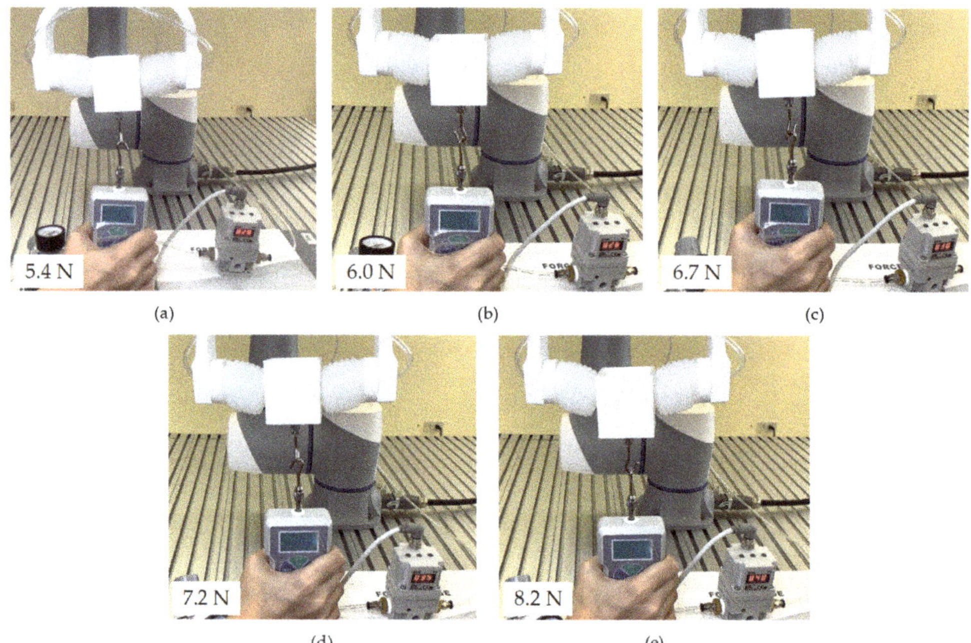

Figure 15. Responses of the pulled gripper at different air pressure levels of (**a**) 20 kPa, (**b**) 25 kPa, (**c**) 30 kPa, (**d**) 35 kPa, and (**e**) 40 kPa.

Figure 16 provides a comparison of the average pulling force exerted on different object shapes. The cube shape exhibited the highest gripping force (8.16 ± 0.30 N), followed by the cylindrical shape (7.14 ± 0.80 N) and the spherical shape (5.18 ± 0.08 N), all tested at 40 kPa. This difference in pulling force can be attributed to the variation in the contact surface area between the gripper and the objects. The cube shape offered a larger contact surface area, resulting in a stronger grip. The pulling force response to pressure changes was linear for each object shape, with the cube and cylinder objects exhibiting a higher rate of pulling force response compared to the sphere object. The designed and optimized soft bellow actuator demonstrated its capability to grasp various object shapes with a robust pulling force. Notably, the cost of the soft bellow prototype is remarkably low, amounting to only USD 1.22, while the total price of the robotic gripper is also economical at USD 5.90.

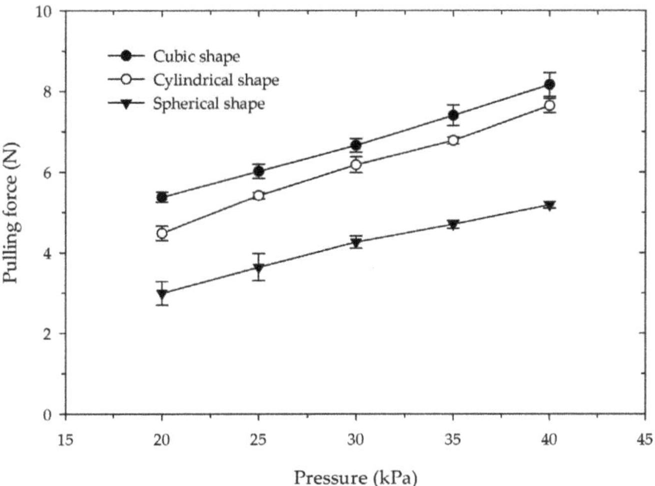

Figure 16. The result of pulling force on different shapes of objects.

7. Conclusions

This study aimed to optimize the design of the soft bellow actuator by considering factors such as wall thickness, number of bellows, and divided ratio of radians, as they significantly affect the deformation and stress response. To achieve this, we employed FEA and a Box–Behnken response surface design, resulting in a design matrix comprising 15 experimental models. Through variance analysis using ANOVA, we identified the key factors influencing the deformation and stress response, with wall thickness emerging as the most significant factor, followed by the divided ratio of radians and the number of bellows. Upon analyzing the design matrix, we determined Trial number 5 to be the optimized design for the soft bellow actuator. This design demonstrated substantial deformation and allowable stress when subjected to maximum air pressure. A comparison between the experimental and simulation results revealed a satisfactory deformation response, with an error rate of approximately 11.46% at high pressure, or 2.18 mm less than the optimized simulation model.

Additionally, we evaluated the pulling-force performance of the validated soft bellow actuator model by integrating two actuators into a rigid gripper structure. The experimental results demonstrated the capability of the soft bellow gripper to grasp objects of various shapes, achieving a maximum holding force of 8.16 N. Consequently, the optimized design and regression model of the soft bellow actuator can be employed in numerical models to assess the suitability of the soft gripper. Additionally, the fabrication process is simplified with the use of a single air chamber, and the displacement of the actuator can be easily controlled by adjusting the pneumatic pressure. Furthermore, the design of the soft bellow gripper offers the flexibility to be integrated with different rigid frames, enabling efficient object grasping in a wide range of applications. This design's versatility allows for its adaptation to various scenarios, thereby enhancing its potential for broader implementation.

Author Contributions: Conceptualization and methodology, J.A. and P.S.; formal analysis and investigation, J.A. and N.V.; writing—original draft preparation, J.A.; writing—review and editing, A.B. and P.S.; supervision, J.A. and P.S. All authors have read and agreed to the published version of the manuscript.

Funding: This work was supported by Research and Development Office, Prince of Songkla University, Thailand under Grant ENG6502013S.

Data Availability Statement: Not applicable.

Conflicts of Interest: The authors declare no conflict of interest.

References

1. Samadikhoshkho, Z.; Zareinia, K.; Janabi-Sharifi, F. A Brief Review on Robotic Grippers Classifications. In Proceedings of the 2019 IEEE Canadian Conference of Electrical and Computer Engineering (CCECE), Edmonton, AB, Canada, 5–8 May 2019; pp. 1–4.
2. Li, M.; Pal, A.; Aghakhani, A.; Pena-Francesch, A.; Sitti, M. Soft Actuators for Real-World Applications. *Nat. Rev. Mater.* **2022**, *7*, 235–249. [CrossRef]
3. Garcia Rubiales, F.J.; Ramon Soria, P.; Arrue, B.C.; Ollero, A. Soft-Tentacle Gripper for Pipe Crawling to Inspect Industrial Facilities Using Uavs. *Sensors* **2021**, *21*, 4142. [CrossRef] [PubMed]
4. Morikage, O.; Wang, Z.; Hirai, S. Multi-Fingered Soft Gripper Driven by Bellows Actuator for Handling Food Materials. In Proceedings of the 2021 27th International Conference on Mechatronics and Machine Vision in Practice (M2VIP), Shanghai, China, 26–28 November 2021; pp. 363–368.
5. Hu, Q.; Dong, E.; Sun, D. Soft Gripper Design Based on the Integration of Flat Dry Adhesive, Soft Actuator, and Microspine. *IEEE Trans. Robot.* **2021**, *37*, 1065–1080. [CrossRef]
6. Pan, M.; Yuan, C.; Liang, X.; Dong, T.; Liu, T.; Zhang, J.; Zou, J.; Yang, H.; Bowen, C. Soft Actuators and Robotic Devices for Rehabilitation and Assistance. *Adv. Intell. Syst.* **2022**, *4*, 2100140. [CrossRef]
7. Heung, H.L.; Tang, Z.Q.; Shi, X.Q.; Tong, K.Y.; Li, Z. Soft Rehabilitation Actuator with Integrated Post-Stroke Finger Spasticity Evaluation. *Front. Bioeng. Biotechnol.* **2020**, *8*, 111. [CrossRef]
8. Sun, Z.; Guo, Z.; Tang, W. Design of Wearable Hand Rehabilitation Glove with Soft Hoop-Reinforced Pneumatic Actuator. *J. Cent. South Univ.* **2019**, *26*, 106–119. [CrossRef]
9. Polygerinos, P.; Wang, Z.; Galloway, K.C.; Wood, R.J.; Walsh, C.J. Soft Robotic Glove for Combined Assistance and At-Home Rehabilitation. *Robot. Auton. Syst.* **2015**, *73*, 135–143. [CrossRef]
10. Navas, E.; Fernández, R.; Armada, M.; Gonzalez-de-Santos, P. Diaphragm-Type Pneumatic-Driven Soft Grippers for Precision Harvesting. *Agronomy* **2021**, *11*, 1727. [CrossRef]
11. Ranasinghe, H.N.; Kawshan, C.; Himaruwan, S.; Kulasekera, A.L.; Dassanayake, P. Soft Pneumatic Grippers for Reducing Fruit Damage During Strawberry Harvesting. In Proceedings of the 2022 Moratuwa Engineering Research Conference (MERCon), Moratuwa, Sri Lanka, 27–29 July 2022; pp. 1–6.
12. Navas, E.; Dworak, V.; Weltzien, C.; Fernández, R.; Shokrian Zeini, M.; Käthner, J.; Shamshiri, R. An Approach to the Automation of Blueberry Harvesting Using Soft Robotics. *43. GIL-Jahrestag. Resiliente Agri-Food-Syst.* 2023, pp. 435–440. Available online: https://dl.gi.de/server/api/core/bitstreams/9cf3b7f2-150d-4c51-951e-b8b40390d9f3/content (accessed on 10 June 2023).
13. Zaidi, S.; Maselli, M.; Laschi, C.; Cianchetti, M. Actuation Technologies for Soft Robot Grippers and Manipulators: A Review. *Curr. Robot. Rep.* **2021**, *2*, 355–369. [CrossRef]
14. Chen, R.; Zhang, C.; Sun, Y.; Yu, T.; Shen, X.-M.; Yuan, Z.-A.; Guo, J.-L. A Paper Fortune Teller-Inspired Reconfigurable Soft Pneumatic Gripper. *Smart Mater. Struct.* **2021**, *30*, 045002. [CrossRef]
15. Walker, J.; Zidek, T.; Harbel, C.; Yoon, S.; Strickland, F.S.; Kumar, S.; Shin, M. Soft Robotics: A Review of Recent Developments of Pneumatic Soft Actuators. *Actuators* **2020**, *9*, 3. [CrossRef]
16. Wang, Z.; Or, K.; Hirai, S. A Dual-Mode Soft Gripper for Food Packaging. *Robot. Auton. Syst.* **2020**, *125*, 103427. [CrossRef]
17. Mosadegh, B.; Polygerinos, P.; Keplinger, C.; Wennstedt, S.; Shepherd, R.F.; Gupta, U.; Shim, J.; Bertoldi, K.; Walsh, C.J.; Whitesides, G.M. Pneumatic Networks for Soft Robotics That Actuate Rapidly. *Adv. Funct. Mater.* **2014**, *24*, 2163–2170. [CrossRef]
18. Auysakul, J.; Vittayaphadung, N.; Gonsrang, S.; Smithmaitrie, P. Bending Angle Effect of TheCross-Section Ratio for a Soft Pneumatic Actuator. *Int. J. Mech. Eng. Robot. Res.* **2020**, *9*, 366–370. [CrossRef]
19. Zhang, X.; Yu, S.; Dai, J.; Oseyemi, A.E.; Liu, L.; Du, N.; Lv, F. A Modular Soft Gripper with Combined Pneu-Net Actuators. *Actuators* **2023**, *12*, 172. [CrossRef]
20. Cheng, P.; Ye, Y.; Yan, B.; Lu, Y.; Wu, C. Eccentric High-Force Soft Pneumatic Bending Actuator for Finger-Type Soft Grippers. *J. Mech. Robot.* **2022**, *14*, 060908. [CrossRef]
21. Ye, Y.; Cheng, P.; Yan, B.; Lu, Y.; Wu, C. Design of a Novel Soft Pneumatic Gripper with Variable Gripping Size and Mode. *J. Intell. Robot Syst.* **2022**, *106*, 5. [CrossRef]
22. Polygerinos, P.; Wang, Z.; Overvelde, J.T.; Galloway, K.C.; Wood, R.J.; Bertoldi, K.; Walsh, C.J. Modeling of Soft Fiber-Reinforced Bending Actuators. *IEEE Trans. Robot.* **2015**, *31*, 778–789. [CrossRef]
23. Joseph, T.; Baldwin, S.; Guan, L.; Brett, J.; Howard, D. The Jamming Donut: A Free-Space Gripper Based on Granular Jamming. *arXiv* **2022**, arXiv:2212.06485.
24. Gao, G.; Chang, C.M.; Gerez, L.; Liarokapis, M. A Pneumatically Driven, Disposable, Soft Robotic Gripper Equipped with Multi-Stage, Retractable, Telescopic Fingers. *IEEE Trans. Med. Robot. Bionics* **2021**, *3*, 573–582. [CrossRef]
25. Guo, N.; Sun, Z.; Wang, X.; Yeung, E.H.K.; To, M.K.T.; Li, X.; Hu, Y. Simulation Analysis for Optimal Design of Pneumatic Bellow Actuators for Soft-Robotic Glove. *Biocybern. Biomed. Eng.* **2020**, *40*, 1359–1368. [CrossRef]
26. Cao, M.; Zhu, J.; Fu, H.; Loic, H.Y.F. Response Surface Design of Bellows Parameters with Negative Pressure Shrinkage Performance. *Int. J. Interact. Des. Manuf. (IJIDeM)* **2022**, *16*, 1041–1052. [CrossRef]

27. Müller, D.; Sawodny, O. Modeling the Soft Bellows of the Bionic Soft Arm. *IFAC-PapersOnLine* **2022**, *55*, 229–234. [CrossRef]
28. Dean, A.; Voss, D.; Draguljić, D. Response Surface Methodology. In *Design and Analysis of Experiments*; Dean, A., Voss, D., Draguljić, D., Eds.; Texts in Statistics; Springer International Publishing: Cham, Switzerland, 2017; pp. 565–614. ISBN 978-3-319-52250-0.
29. Kleijnen, J.P.C. Response Surface Methodology. In *Handbook of Simulation Optimization*; Fu, M.C., Ed.; International Series in Operations Research & Management Science; Springer: New York, NY, USA, 2015; pp. 81–104. ISBN 978-1-4939-1384-8.
30. Khuri, A.I.; Mukhopadhyay, S. Response Surface Methodology. *WIREs Comput. Stat.* **2010**, *2*, 128–149. [CrossRef]
31. Honarpardaz, M. A Methodology for Design and Simulation of Soft Grippers. *Simulation* **2021**, *97*, 779–791. [CrossRef]
32. Chen, F.; Xu, W.; Zhang, H.; Wang, Y.; Cao, J.; Wang, M.Y.; Ren, H.; Zhu, J.; Zhang, Y.F. Topology Optimized Design, Fabrication, and Characterization of a Soft Cable-Driven Gripper. *IEEE Robot. Autom. Lett.* **2018**, *3*, 2463–2470. [CrossRef]
33. Tawk, C.; Alici, G. Finite Element Modeling in the Design Process of 3D Printed Pneumatic Soft Actuators and Sensors. *Robotics* **2020**, *9*, 52. [CrossRef]
34. Guo, L.; Li, K.; Cheng, G.; Zhang, Z.; Xu, C.; Ding, J. Design and Experiments of Pneumatic Soft Actuators. *Robotica* **2021**, *39*, 1806–1815. [CrossRef]
35. Marechal, L.; Balland, P.; Lindenroth, L.; Petrou, F.; Kontovounisios, C.; Bello, F. Toward a Common Framework and Database of Materials for Soft Robotics. *Soft Robot.* **2021**, *8*, 284–297. [CrossRef]
36. Szurgott, P.; Jarzębski, Ł. Selection of a Hyper-Elastic Material Model-a Case Study for a Polyurethane Component. *Lat. Am. J. Solids Struct.* **2019**, *16*, 1–16. [CrossRef]

 *actuators* 

Article

Characteristic Analysis of Heterochiral TCP Muscle as a Extensile Actuator for Soft Robotics Applications

Beau Ragland and Lianjun Wu *

Department of Manufacturing Engineering, Georgia Southern University, Statesboro, GA 30461, USA; beauragland@gmail.com
* Correspondence: lwu@georgiasouthern.edu

Abstract: A soft actuator is an essential component in a soft robot that enables it to perform complex movements by combining different fundamental motion modes. One type of soft actuator that has received significant attention is the twisted and coiled polymer artificial muscle (TCP actuator). Despite many recent advancements in TCP actuator research, its use as an extensile actuator is less common in the literature. This works introduces the concept of using TCP actuators as thermal-driven extensile actuators for robotics applications. The low-profile actuator can be easily fabricated to offer two unique deformation capabilities. Results from the characterization indicate that extensile actuators, made with various rod diameters and under different load conditions, display remarkable elongation deformation. Additionally, a proof-of-concept soft-earthworm robot was developed to showcase the potential application of the extensile actuator and to demonstrate the benefits of combining different types of motion modes.

Keywords: artificial muscle; twisted and coiled polymer artificial muscle; extensile actuator; soft robotics

Citation: Ragland, B.; Wu, L. Characteristic Analysis of Heterochiral TCP Muscle as a Extensile Actuator for Soft Robotics Applications. *Actuators* **2023**, *12*, 189. https://doi.org/10.3390/act12050189

Academic Editor: Hamed Rahimi Nohooji

Received: 27 March 2023
Revised: 17 April 2023
Accepted: 25 April 2023
Published: 28 April 2023

1. Introduction

Soft robotics is an emerging field of robotics that focuses on creating highly deformable, flexible robots. An essential component of soft robots is the soft actuator, which enables soft robots to perform intricate movements and interact with their surroundings. Soft actuators can produce a range of basic actuation responses such as contraction, expansion, rotation, and bending, allowing soft robots to achieve a variety of complex behaviors.

The specific type of motion generated by a soft actuator depends on its design and the materials used. For instance, shape memory alloy (SMA) contracts in response to heat, which has been leveraged by researchers to produce mechanical work [1,2]. Pneumatic artificial muscles (PAMs) and hydraulic actuators contract in response to changes in air or hydraulic pressure, respectively [3–5]. Torsional actuation originating from structural twist behavior can be found in twisted nano-/macrofiber yarns [6,7], twisted and coiled polymer artificial muscle [8–11], spider dragline silk [12], graphene oxide [13] and SMA fibers [14,15]. Bending motion can be found in ionic polymer–metal composites (IPMCs) [16,17], which use the movement of ions under the influence of an electric field. Dielectric elastomer actuators can also produce a bending motion in response to an applied electric field [18,19]. By conducting structural design and combining multiple basic actuation responses, complex motions such as twisting [20,21], curling [22,23], and folding [24,25] can also be achieved in the aforementioned actuators. However, the concept of deploying elongation as a driving force is less common in the literature. DEA artificial muscles with a certain degree of anisotropy can exhibit an expanding motion [26], forming antagonistic soft actuator systems to mimic biological systems [27,28]. However, the high operating voltage of dielectric elastomer actuators presents a significant barrier to their extensive adoption in soft robotics [29,30]. For pneumatic actuators, different jamming structures [31,32] and alterations in the braid angle of PAMs [33,34] can enable elongation motion. However, high noise levels and bulky forms may make them unsuitable for certain applications where a quiet operation or compact size is important.

In this work, we reported twisted and coiled polymer (TCP) muscle as an extensile actuator capable of performing mechanical work and demonstrated the possibility of creating soft robots for the first time. The research presented in this paper builds upon the work conducted by Beau in his thesis [35]. Recently, TCP actuators have attracted wide attention in various applications, including soft robots [36,37], soft grippers [38,39], robotic hand/prosthetic hands [40,41], orthotic devices [42,43], sensors [44,45], musculoskeletal systems [46], healing composites [47], and energy harvesters [48], due to their high strain and force generation, low weight and low-profile nature. These applications highlight that TCP actuators are mainly used as torsional or contractile actuators.

Despite many recent advancements in TCP actuator research, their use as extensile actuators is less common in the literature. This is the first study to use TCP muscle as an extensile actuator in soft robotics. The main contributions of this paper are three-fold. First, this work proposes a thermal-driven extensile actuator for soft robotics, which broadens the range of motion modes available for this type of actuator. It provides an additional novel solution for powering soft robots. Second, this work presents a detailed experimental characterization of TCP muscle as an extensile actuator, showing its potential to perform mechanical work. Finally, the paper demonstrates the feasibility of applying an extensile actuator in soft-earthworm robots, highlighting their potential for use in soft robotics applications.

The rest of this article is organized as follows. We first introduce the fabrication method for this type of heterochiral artificial muscles, followed by the characterization method in Section 2. In Section 3, the properties and performance of the actuator are tested, followed by a proof-of-concept demonstration in soft robotics. Finally, Section 4 concludes this article.

2. Materials and Methods

2.1. Heterochiral Artificial Muscle Fabrication

We first fabricated the heterochiral artificial muscle using a nylon 6 monofilament fishing line fiber (Eagle Claw monofilament 80 lb-test fishing line with a diameter of 0.8 mm). The fabrication process for this type of heterochiral actuator followed the four major steps that were reported in our previous work [46]: twist insertion (Figure 1c), resistance wire wrapping (Figure 1d), mandrel coiling (Figure 1e) and thermal annealing.

To achieve twist insertion, we twisted the 0.8-mm diameter nylon 6 fiber in the Z direction (clockwise) using the setup in Figure 1c until the first coil formation was observed along the fiber. A 500-g deadweight was used to keep the fishing line fiber under tension during the twist insertion process. For resistance wire wrapping, we rapped a thin nickel-chromium wire with a diameter of 0.076 mm around the twisted fiber using the setup in Figure 1d. The conductive wire pool was placed on the pool holder. The pool holder was mounted to the linear actuator. The linear actuator traveled along the axis direction of the twisted fiber. Both stepper motors (M3) rotated in the opposite direction to prevent any additional twist from being added to the twisted fiber. With the combination of linear movement of the linear actuator and rotational movement of the stepper motors (M3), the conductive wire was woven around the twisted line. To ensure the conducting wire does not become bunched up within the feeding mechanism, we attached a small 10 g weight to keep the wire taut. Next, we performed mandrel coiling by wrapping the fishing line onto the rod in the S direction (counterclockwise). The major difference in this work from the previous work is that the actuator is oppositely twisted and coiled to form a heterochiral structure. The term "heterochiral" refers to the opposite chirality or handedness of the TCP muscles in their structure compared to the direction of their twisting or coiling. The final step was thermal annealing. We placed the coiled artificial muscle, with both ends tethered, in an oven (Memmert Universal Oven) where the temperature was set to 180 degrees Celsius for 90 min. The same annealing process was used for all the muscles used in this work. The thermal expansion induced untwist of the fiber produced the untwist of the helical structure which resulted in the muscle elongation [11].

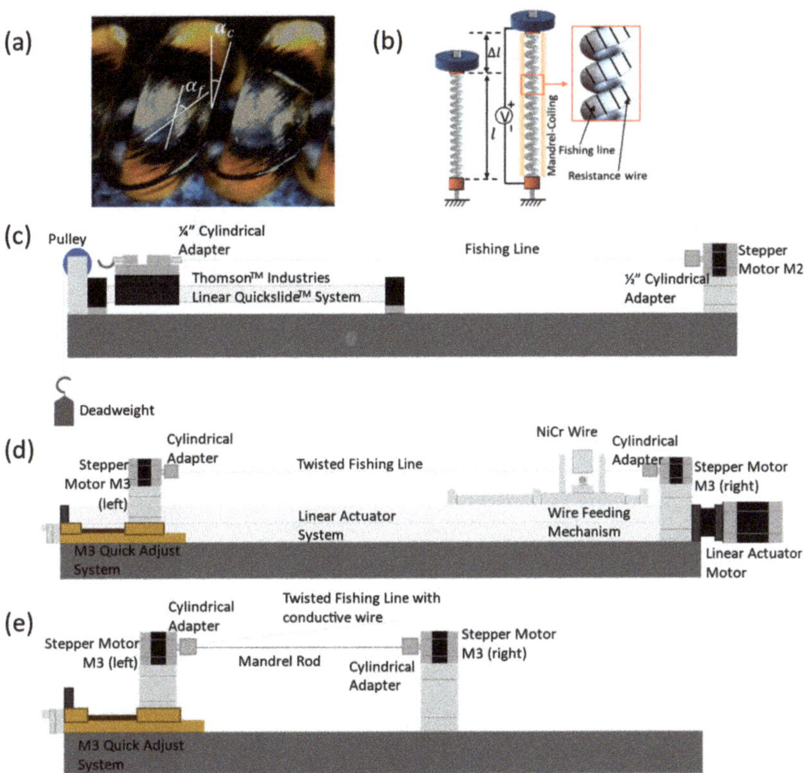

Figure 1. (**a**) The extensile artificial muscle (heterochiral structure) with a fiber bias angle (α_f = 37°) and coil bias angle (α_c = 16°) measured from a microscope. (**b**) Schematic illustration of elongational deformation realized by heterochiral coiled artificial muscle. (**c**) Inserting twist into the pristine fishing line. (**d**) Incorporating the nickel-chromium wire (resistance wire) into the muscle by wrapping the resistance wire around the twisted fiber. (**e**) Forming the mandrel-coiling muscle from the twisted fiber with resistance wire as made in (**d**) by wrapping it around the rod via the rotation of the stepper motor.

Figure 1a shows a microscopic image of the heterochiral TCP actuator. The image was taken using a microscope (HAYEARTM HY-150X) and processed with the proprietary microscope image processing software to measure the fiber bias angle α_f and coil bias angle α_c. The fiber bias angle (an angle between the fiber axis and its initial orientation) resulted from the applied torque during the twist insertion step, where the disordered molecular chains became aligned. The initial coil bias angle (an angle between the fiber and the coil's cross-section) resulted from the mandrel-coiling step. Figure 1b illustrates that upon receiving a thermal stimulus, the actuator can elongate along its axis direction to lift a deadweight and perform mechanical work. The displacement ratio (tensile actuation strain ε) is expressed as a percentage and calculated as $\varepsilon = (\Delta l / l) * 100$, where Δl and l are the muscle deformation displacement after actuation and its original lengths before actuation.

2.2. Characterization Method

The lengthwise elongation of the extensile TCP actuator upon heating with varying mandrel rod diameters and deadweights was characterized using an experimental setup, as depicted in Figure 2. A laser displacement sensor (Keyence IL-300) was used to measure the elongation displacement, while a power supply (BK Precision 9206) was used to program and define the power input waveforms. We intend to control the current instead of the voltage, as different lengths of muscle require different voltages. A deadweight was placed on top of a 3D-printed plate mounted to the low-friction linear ball slide, which could freely move up and down. It is very challenging to have an accurate and reliable measurement of temperature due to the small diameter of the fiber. In this study, we tried our best to use thermocouples for temperature measurement. Even though the temperature might not be accurate, we relied on this to monitor the trend of the muscles' temperature to prevent overheating-induced damage. To be specific, two type-E thermocouples from Omega Engineering were used to monitor the muscle's temperature. The data were acquired using a National Instrument DAQ.

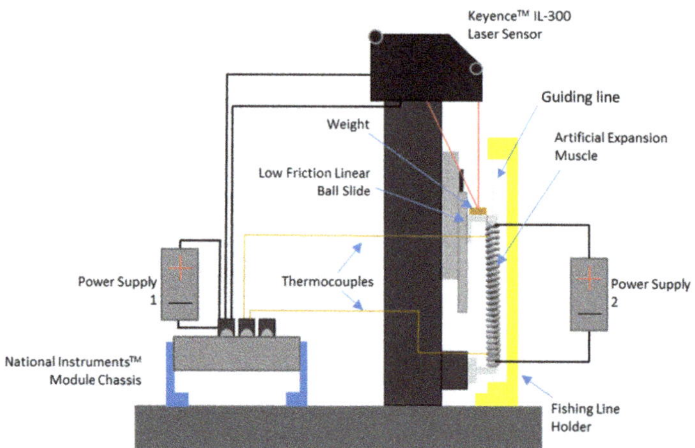

Figure 2. Schematic diagram of the experimental setup to characterize the extensile TCP actuator.

Due to the tendency of the extensile TCP actuator to buckle, it is necessary to constrain the out-of-plane deflection during testing. A guiding line (a thin non-twisted fishing line) was run through the center of the muscle, which was fastened to the yellow fishing line holder. This guiding line kept the muscle upright while remaining electrically insulated to avoid interfering with the electrical current flowing through the muscle. Before each muscle was tested, it was trained via 10 cycles of heating and cooling to eliminate the lonely stroke while testing. Based on the results from the preliminary testing, a regulated constant current of 0.20 A and a 60-s period with a 25% duty cycle (15 s on vs. 45 s off) were used for the characterization test. This period was repeated five times for a total of 300 s. Data were acquired via a LabVIEW program and were plotted into graphs via a MATLAB program.

3. Results and Discussion

3.1. Spring Index

We first investigated the effect of the spring index on the extensile actuator's performance. The spring index is a ratio of the mean coil diameter to the fiber diameter, and it determines the final coil diameter. Due to the intrinsic structural nature of the coiled muscle, the mean coil diameter is determined by the diameter of the rod as well as the fiber diameter. The rod was used in the muscle fabrication to form a mandrel-coiled structure. In this study, the extensile actuators were fabricated with varying mandrel rod diameters of 0.035″, 0.041″, and 0.055″. The spring index is directly related to the mandrel coil diameter.

The spring index of the muscle made from a 0.035″ rod diameter was 2.98; the spring index of the muscle made from a 0.041″ rod diameter was 3.23; and the spring index of the muscle made from a 0.055″ rod diameter was 3.63. A regulated constant current of 0.20 A was applied to each muscle, and no deadweight was applied for this experiment. The loaded lengths were 73.4 mm, 73.2 mm, and 68.3 mm, respectively.

Figure 3a,b show the fabricated actuators' ability to deform in the direction of the axis. It is obvious when viewing Figure 3a,b that the muscle made from the 0.055″ rod diameter with the largest spring index produced the largest absolute displacement with a peak value of 35.3 mm as well as the largest displacement ratio reaching up to 54%. The larger spring index contributed to the actuator's ability to deform along the axis, resulting in greater displacement under a given load. After completing the cooling cycle, the preliminary temperature results showed that the muscle temperature ranged from 42 °C to 47 °C. The experiment's cooling time setting may have been insufficient for the muscle made from the 0.055″ rod diameter to fully return to its original position. Therefore, additional cooling time or the use of active cooling may be necessary to achieve complete recovery. Overall, the data indicate a distinct pattern where the increase in the diameter of the rod results in a larger spring index, which contributes to a greater deformation.

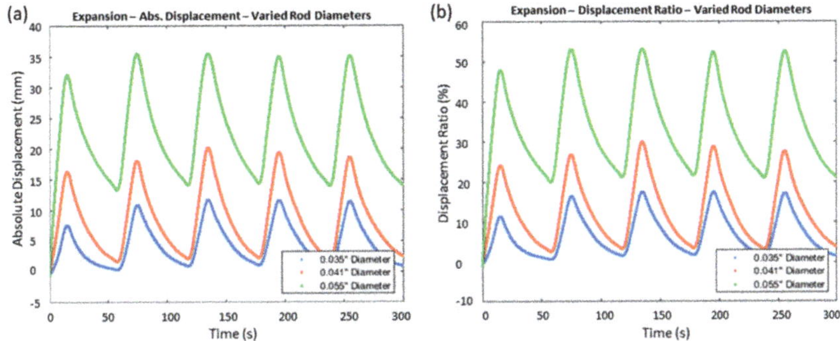

Figure 3. The elongation performance of the muscles made from various rod diameters obtained at a regulated current of 0.2 A. The absolute displacement (**a**) and the displacement ratio (**b**) of an extensile actuator with different spring indexes made from different varying rod diameters.

3.2. Various Load Conditions

In addition to investigating the effect of the spring index, we also examined the deformation of the actuator under various load conditions using the muscle made from a 0.055″ rod diameter, which exhibited desirable deformation properties. To ensure accurate and consistent results, we performed a training phase for at least 10 cycles on each muscle by applying a regulated constant current of 0.20 A at various load conditions. Three deadweights of 20 g, 40 g, and 60 g were used, and the loaded lengths were 67.6 mm, 67.4 mm, and 67.3 mm, respectively.

Our findings, depicted in Figure 4a,b, show that the maximum peak absolute displacement decreased as the deadweight increased. Specifically, the muscle exhibited a maximum peak absolute displacement of 32.6 mm under a 20 g deadweight, while the deadweights of 40 g and 60 g resulted in a maximum absolute displacement of 26.1 mm and 14.6 mm, respectively. The maximum displacement ratio achieved for each weight was 48.2% for 20 g, 38.9% for 40 g, and 21.8% for 60 g. Based on the preliminary temperature results, the residual temperature was between 50 °C and 57 °C, which we believe may negatively impact the life cycle performance due to the accumulative effect of the temperature on the system over time. Therefore, for the endurance test, we reduced the test current to 0.18 A to extend its performance longevity. Furthermore, we observed that the greater the compressive load, the more it facilitated the muscle's return motion, indicating the actuator's resilience to external pressure. Overall, the results suggest that the extensile actuator has promising applications in load-bearing and deformation-sensitive tasks.

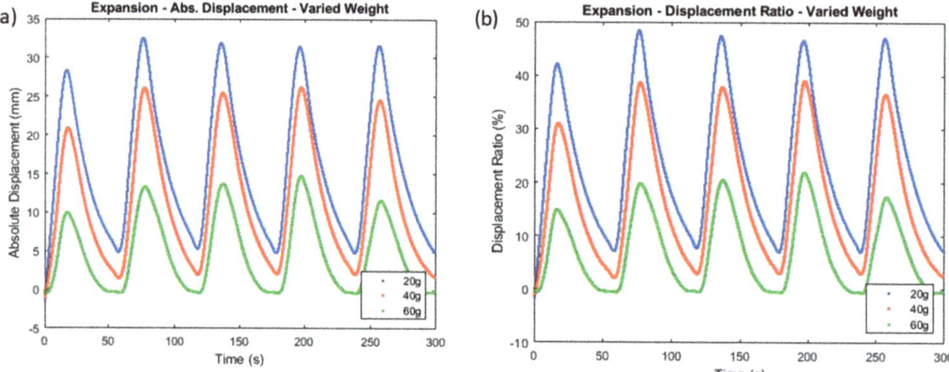

Figure 4. The elongation performance of the muscles at various load conditions obtained at a regulated current of 0.2 A. The absolute displacement (**a**) and the displacement ratio (**b**) of the extensile actuator subjected to different deadweights.

3.3. Endurance Test

The endurance test was performed to determine how long the actuator can perform at an acceptable level without failure or degradation. A muscle made from a 0.055″ rod diameter with a loaded length of 45.1 mm was used with a 20 g deadweight. For the endurance test, we reduced the current to 0.18 A to extend its performance longevity. The muscle was tested at a duty cycle of 25% and a period of 60 s for 1000 cycles continuously in ambient air. Figure 5a shows that after around 80 cycles, the peak displacement ratio dropped from 20% and stabilized around 13% through the rest of the cycles. Figure 5b,c show a subtle visual color change in the muscle starting from cycle 0 until the end of the test. It was observed that the heating wire left some imprints or marks on the surface of the fishing line, which might contribute to the slight degradation of the actuator's performance.

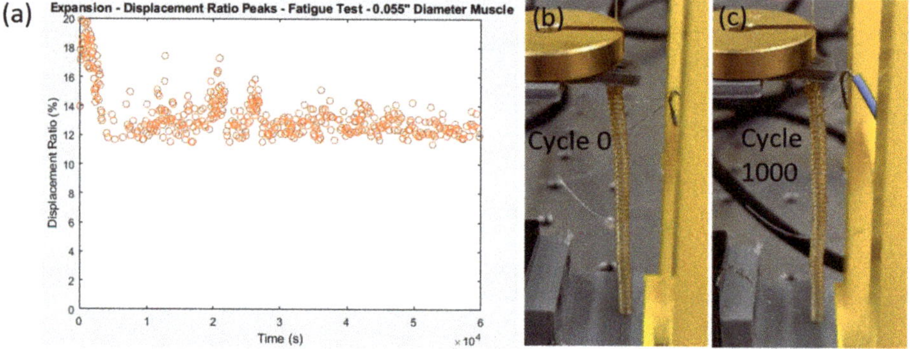

Figure 5. (**a**) Endurance test of the muscle driven by a constant current (0.18 A, duty cycle 25% and a period of 60 s) under a constant 20 g load for 1000 cycles of actuation (total time 60,000 s). The pictures of the muscle at cycle 0 (**b**) and cycle 1000 (**c**) indicate a slight discoloration.

3.4. Application of The Extensile Actuator

To further illustrate the potential application of the extensile actuator and demonstrate the advantages of combining different types of motion modes, we developed a soft-earthworm robot that combines the contraction and elongation capabilities of artificial muscles. The earthworm robot moves by relying on frictional interaction with the surfaces of the vessel it is placed in [49]. The ability to adjust this frictional interaction is a crucial aspect of moving effectively. To showcase this capability, we designed the end caps of the robot with extensile artificial muscles that manipulate frictional forces to assist locomotion.

In this study, as depicted in Figure 6, the silicone shell was fabricated using a Hyrel modular 3D printer (Hydra 16 A). The shell was manufactured first by producing a hollow mold made of water-soluble PVA as shown in Figure 6a. The printer bed was covered with a layer of painter's tape to allow for better adhesion of the print to the build plate as shown in Figure 6b. Figure 6c shows that after the PVA mold was finished, a static mixing head connected to two reservoir heads hosting silicone parts A and B, respectively, was then used to fill the mold with Smooth-On Ecoflex 00–30 silicone. The static mixing head permits the mixing in a 1:1 ratio between parts A and B. After the printing was finished, we waited for 4 h before taking the entire build-out of the printer and placing it into a Vevor 26–30 L Digital Ultrasonic Cleaner at 35 °C. Figure 6d shows the final fabricated silicone shell.

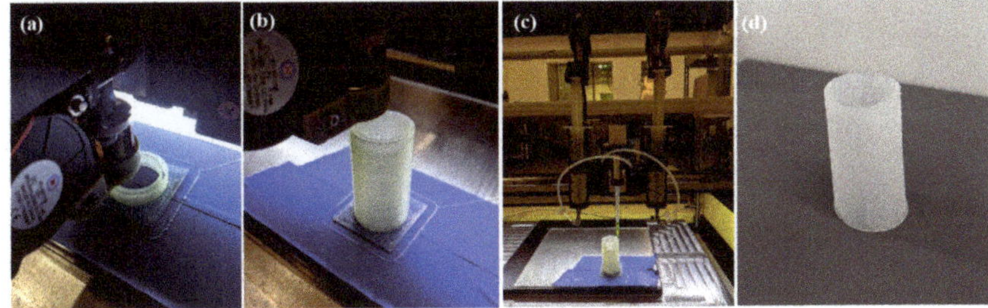

Figure 6. (**a**) The first few layers of printing PVA mold. (**b**) The completed printing of PVA mold. (**c**) A static mixing head that dynamically mixes silicones at a 1:1 ratio dispenses silicone into a PVA mold (**d**) Finished silicone shell.

The end caps of the robot with extensile artificial muscles are mainly used for providing two friction modes to assist locomotion. Figure 7a shows the end cap in a high friction mode, with the U-shaped variable friction piece in the up position (actuators were not actuated); in Figure 7b, the actuated muscles pushed the U-shaped variable friction piece out of the end cap, reducing the contact area and minimizing friction with the surface. When reduced friction is no longer needed, the muscles naturally cool and retract, pulling the friction piece back up into the body of the end cap. Rubber tips were added at the base of the design to prevent slippage. Figure 7c shows the prototype of the earthworm, which uses a total of seven muscles. Due to the tendency of the extensile TCP actuator to buckle, it is very challenging to constrain the out-of-plane deflection during motion. Therefore, contractile muscles were used in the middle long section and extensile muscles were used in the end sections. Three contractile muscles were fabricated using the same method reported in our previous work [46]. These muscles were placed equidistant from each other in the middle section of the robot, and a long spring was included in the center of the robot to form an antagonist pair. When the contractile muscles in the middle section are activated, the length of the prototype shortens. However, the central spring enabled the robot to return to its original length when heat was no longer applied to the contraction muscles. Four extensile muscles were housed in both ends of the robot to alter frictional interaction. The contraction muscles were attached to the end caps with Loctite Instant Mix Epoxy. The silicone shell was attached to the end cap around the artificial muscles via the same epoxy.

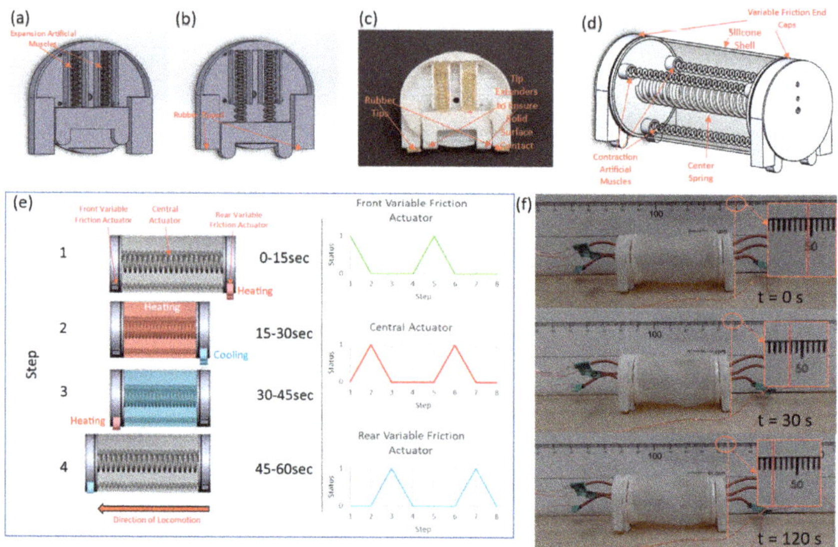

Figure 7. A variable friction end cup powered by the extensile actuator in high-friction mode (**a**) and low-friction mode (**b**). (**c**) The prototype of the variable friction end cup. (**d**) The design concept of the earthworm robot. (**e**) The breakdown of the actuation sequence is in four steps. (**f**) Photographic sequence with time stamps showing the horizontal locomotion of the earthworm (magnified views of the scale reading are inset).

To move forward, the contraction muscles were periodically activated in conjunction with the extensile muscles located within the variable friction endcaps. This allowed the robot to move forward with only one segment, and the movement was broken into four repeating steps as shown in Figure 7e. In step 1 (0–15 s), the front variable friction actuators were powered on to create a smaller contact area in the rear. In step 2 (15–30 s), the central actuators were powered on to pull the rear-end cap forward. In step 3 (30–45 s), the front variable friction actuators were powered on to allow the front-end cap to be pushed forward. In step 4 (45–60 s), the front variable friction actuators were powered off to prepare for the next cycle of motion. The entire sequence of movement was programmed and controlled via a PLC (Triangle Research Fx2424 PLC, White Rock, BC, Canada), and the locomotion of the earthworm robot was recorded using a web camera.

Figure 7f shows the locomotion of the robot, with the central contraction muscle actuated at 0.36 A and the extensile muscle actuated at 0.20 A. The top picture in Figure 7f shows the initial state at 0 s; the middle picture shows that peak deformation occurred in the central actuator at 30 s; the bottom picture shows the final state at 120 s. The maximum axial move distance is 3.7 mm. One of the limitations of the thermal-driven actuator is the efficiency; the maximum energy conversion efficiency during contraction for homochiral artificial muscle that can contract is less than 1.08% [11]. Although the proposed soft robot has limitations in terms of movement speed and operation frequency, the combination of multiple motion modes shows the promising potential application of extensile actuators. To overcome these limitations, future work could focus on improving the design of the actuating system, incorporating active cooling for faster actuation, exploring cascading structures composed of multiple units for complex robotic systems, or designing new control strategies that enable faster and more precise movements.

4. Conclusions

This paper described the fabrication and characterization of a heterochiral extensile muscle that had a unique capability for elongation compared to other widely reported homochiral contractile muscles. Upon receiving Joule heat, the muscle's fiber untwists due to thermal expansion, resulting in the untwisting of the helical structure and causing muscle elongation. The results show that the extensile actuator's performance is affected by the spring index, which is directly related to the mandrel coil diameter. Increasing the mandrel rod's diameter and hence the spring index leads to greater deformation and displacement of the actuator. However, additional cooling time or the use of active cooling may be necessary to achieve complete recovery.

In addition to characterizing the actuator, we also tested the deformation of the actuator under various load conditions. The actuator made from a 0.055″ diameter rod was tested under different deadweight conditions and found to be inversely related to the load applied to it. Additionally, the greater the compressive load, the more it facilitated the actuator's return motion.

The study also explored the actuator's potential as an earthworm soft robot drive system. The actuator can deliver not only contraction but also elongation motion, making it suitable for many robotic applications where elongation is desired. These actuators are expected to play a prominent role in tasks that require low-cost, silent operation, and high-performance and multi-motion modes. The combination of these basic actuation motion modes also offers the potential to produce other complex motions.

Author Contributions: B.R.: conceptualization, data acquisition, visualization, investigation, writing. L.W.: conceptualization, supervision, writing, funding acquisition. All authors have read and agreed to the published version of the manuscript.

Funding: This research received no external funding.

Data Availability Statement: The data that support the findings of this study are available from the corresponding author (lwu@georgiasouthern.edu) upon reasonable request.

Acknowledgments: We acknowledge the financial support by the Startup from Georgia Southern University. Most of the materials contained in the paper were previously published in Beau Ragland's master thesis https://digitalcommons.georgiasouthern.edu/cgi/viewcontent.cgi?article=3539&context=etd (accessed on 27 March 2023).

Conflicts of Interest: The authors declare no conflict of interest.

References

1. Tadesse, Y.; Hong, D.; Priya, S. Twelve Degree of Freedom Baby Humanoid Head Using Shape Memory Alloy Actuators. *J. Mech. Robot.* **2011**, *3*, 011008. [CrossRef]
2. Huang, X.; Kumar, K.; Jawed, M.K.; Nasab, A.M.; Ye, Z.; Shan, W.; Majidi, C. Highly Dynamic Shape Memory Alloy Actuator for Fast Moving Soft Robots. *Adv. Mater. Technol.* **2019**, *4*, 1800540. [CrossRef]
3. Ohta, P.; Valle, L.; King, J.; Low, K.; Yi, J.; Atkeson, C.G.; Park, Y.-L. Design of a Lightweight Soft Robotic Arm Using Pneumatic Artificial Muscles and Inflatable Sleeves. *Soft Robot.* **2018**, *5*, 204–215. [CrossRef] [PubMed]
4. Lee, J.-G.; Rodrigue, H. Origami-Based Vacuum Pneumatic Artificial Muscles with Large Contraction Ratios. *Soft Robot.* **2018**, *6*, 109–117. [CrossRef]
5. Skowrońska, J.; Kosucki, A.; Stawiński, Ł. Overview of Materials Used for the Basic Elements of Hydraulic Actuators and Sealing Systems and Their Surfaces Modification Methods. *Materials* **2021**, *14*, 1422. [CrossRef] [PubMed]
6. Lima, M.D.; Li, N.; de Andrade, M.J.; Fang, S.; Oh, J.; Spinks, G.M.; Kozlov, M.E.; Haines, C.S.; Suh, D.; Foroughi, J.; et al. Electrically, Chemically, and Photonically Powered Torsional and Tensile Actuation of Hybrid Carbon Nanotube Yarn Muscles. *Science* **2012**, *338*, 928–932. [CrossRef]
7. Mu, J.; de Andrade, M.J.; Fang, S.; Wang, X.; Gao, E.; Li, N.; Kim, S.H.; Wang, H.; Hou, C.; Zhang, Q.; et al. Sheath-run artificial muscles. *Science* **2019**, *365*, 150–155. [CrossRef]
8. Kimura, D.; Irisawa, T.; Takagi, K.; Tahara, K.; Sakurai, D.; Watanabe, H.; Takarada, W.; Shioya, M. Mechanism for anisotropic thermal expansion of polyamide fibers. *Sens. Actuators B Chem.* **2021**, *344*, 130262. [CrossRef]
9. Aziz, S.; Naficy, S.; Foroughi, J.; Brown, H.R.; Spinks, G.M. Controlled and scalable torsional actuation of twisted nylon 6 fiber. *J. Polym. Sci. Part B Polym. Phys.* **2016**, *54*, 1278–1286. [CrossRef]

10. Kim, K.; Cho, K.H.; Jung, H.S.; Yang, S.Y.; Kim, Y.; Park, J.H.; Jang, H.; Nam, J.; Koo, J.C.; Moon, H.; et al. Double Helix Twisted and Coiled Soft Actuator from Spandex and Nylon. *Adv. Eng. Mater.* **2018**, *20*, 1800536. [CrossRef]
11. Haines, C.S.; Lima, M.D.; Li, N.; Spinks, G.M.; Foroughi, J.; Madden, J.D.W.; Kim, S.H.; Fang, S.; de Andrade, M.J.; Göktepe, F.; et al. Artificial Muscles from Fishing Line and Sewing Thread. *Science* **2014**, *343*, 868–872. [CrossRef] [PubMed]
12. Liu, D.; Tarakanova, A.; Hsu, C.C.; Yu, M.; Zheng, S.; Yu, L.; Liu, J.; He, Y.; Dunstan, D.J.; Buehler, M.J. Spider dragline silk as torsional actuator driven by humidity. *Sci. Adv.* **2019**, *5*, eaau9183. [CrossRef] [PubMed]
13. Kim, H.; Moon, J.H.; Mun, T.J.; Park, T.G.; Spinks, G.M.; Wallace, G.G.; Kim, S.J. Thermally Responsive Torsional and Tensile Fiber Actuator Based on Graphene Oxide. *ACS Appl. Mater. Interfaces* **2018**, *10*, 32760–32764. [CrossRef] [PubMed]
14. Yang, S.Y.; Kim, K.; Seo, S.; Shin, D.; Park, J.H.; Gong, Y.J.; Choi, H.R. Hybrid Antagonistic System with Coiled Shape Memory Alloy and Twisted and Coiled Polymer Actuator for Lightweight Robotic Arm. *IEEE Robot. Autom. Lett.* **2022**, *7*, 4496–4503. [CrossRef]
15. Shim, J.-E.; Quan, Y.-J.; Wang, W.D.; Rodrigue, H.; Song, S.-H.; Ahn, S.-H. A smart soft actuator using a single shape memory alloy for twisting actuation. *Smart Mater. Struct.* **2015**, *24*, 125033. [CrossRef]
16. Bian, C.; Zhu, Z.; Bai, W.; Chen, H. Highly efficient structure design of bending stacking actuators based on IPMC with large output force. *Smart Mater. Struct.* **2021**, *30*, 075033. [CrossRef]
17. Tamagawa, H.; Watanabe, H.; Sasaki, M. Bending direction change of IPMC by the electrode modification. *Sens. Actuators B Chem.* **2009**, *140*, 542–548. [CrossRef]
18. Guo, Y.; Guo, J.; Liu, L.; Liu, Y.; Leng, J. Bioinspired multimodal soft robot driven by a single dielectric elastomer actuator and two flexible electroadhesive feet. *Extreme Mech. Lett.* **2022**, *53*, 101720. [CrossRef]
19. Franke, M.; Ehrenhofer, A.; Lahiri, S.; Henke, E.-F.M.; Wallmersperger, T.; Richter, A. Dielectric Elastomer Actuator Driven Soft Robotic Structures with Bioinspired Skeletal and Muscular Reinforcement. *Front. Robot. AI* **2020**, *7*, 510757. [CrossRef]
20. Rodrigue, H.; Bhandari, B.; Han, M.-W.; Ahn, S.-H. A shape memory alloy–based soft morphing actuator capable of pure twisting motion. *J. Intell. Mater. Syst. Struct.* **2014**, *26*, 1071–1078. [CrossRef]
21. Jiao, Z.; Ji, C.; Zou, J.; Yang, H.; Pan, M. Vacuum-Powered Soft Pneumatic Twisting Actuators to Empower New Capabilities for Soft Robots. *Adv. Mater. Technol.* **2019**, *4*, 1800429. [CrossRef]
22. Wakimoto, S.; Ogura, K.; Suzumori, K.; Nishioka, Y. Miniature Soft Hand with Curling Rubber Pneumatic Actuators. In Proceedings of the 2009 IEEE International Conference on Robotics and Automation, Kobe, Japan, 12–17 May 2009; IEEE: Piscataway, NJ, USA.
23. Jones, T.J.; Jambon-Puillet, E.; Marthelot, J.; Brun, P.-T. Bubble casting soft robotics. *Nature* **2021**, *599*, 229–233. [CrossRef] [PubMed]
24. Yu, M.; Yang, W.; Yu, Y.; Cheng, X.; Jiao, Z. A Crawling Soft Robot Driven by Pneumatic Foldable Actuators Based on Miura-Ori. *Actuators* **2020**, *9*, 26. [CrossRef]
25. Wu, S.; Ze, Q.; Zhang, R.; Hu, N.; Cheng, Y.; Yang, F.; Zhao, R. Symmetry-Breaking Actuation Mechanism for Soft Robotics and Active Metamaterials. *ACS Appl. Mater. Interfaces* **2019**, *11*, 41649–41658. [CrossRef]
26. Lu, T.; Shi, Z.; Shi, Q.; Wang, T. Bioinspired bicipital muscle with fiber-constrained dielectric elastomer actuator. *Extreme Mech. Lett.* **2016**, *6*, 75–81. [CrossRef]
27. Liu, Y.; Liu, B.; Yin, T.; Xiang, Y.; Zhou, H.; Qu, S. Bistable rotating mechanism based on dielectric elastomer actuator. *Smart Mater. Struct.* **2019**, *29*, 015008. [CrossRef]
28. Vo, V.T.K.; Ang, M.H.; Koh, S.J.A. Maximal Performance of an Antagonistically Coupled Dielectric Elastomer Actuator System. *Soft Robot.* **2020**, *8*, 200–212. [CrossRef]
29. Guo, Y.; Liu, L.; Liu, Y.; Leng, J. Review of Dielectric Elastomer Actuators and Their Applications in Soft Robots. *Adv. Intell. Syst.* **2021**, *3*, 2000282. [CrossRef]
30. Wang, Y.; Ma, X.; Jiang, Y.; Zang, W.; Cao, P.; Tian, M.; Ning, N.; Zhang, L. Dielectric elastomer actuators for artificial muscles: A comprehensive review of soft robot explorations. *Resour. Chem. Mater.* **2022**, *1*, 308–324. [CrossRef]
31. Fracczak, L.; Nowak, M.; Koter, K. Flexible push pneumatic actuator with high elongation. *Sens. Actuators A Phys.* **2021**, *321*, 112578. [CrossRef]
32. Gai, L.-J.; Zong, X.; Huang, J. Enhancing the Tensile-Shaping Stability of Soft Elongation Actuators for Grasping Applications. *IEEE Robot. Autom. Lett.* **2022**, *8*, 600–607. [CrossRef]
33. Pillsbury, E.T.; Wereley, N.M.; Guan, Q. Comparison of contractile and extensile pneumatic artificial muscles. *Smart Mater. Struct.* **2017**, *26*, 095034. [CrossRef]
34. Iwata, K.; Suzumori, K.; Wakimoto, S. Development of contraction and extension artificial muscles with different braid angles and their application to stiffness changeable bending rubber mechanism by their combination. *J. Robot. Mechatron.* **2011**, *23*, 582–588. [CrossRef]
35. Ragland, B. Design and Development of Soft Earthworm Robot Driven by Fibrous Artificial Muscles, in Department of Manufacturing Engineering. Master's Thesis, Georgia Southern University, Statesboro, GA, USA, 2021. Available online: https://digitalcommons.georgiasouthern.edu/cgi/viewcontent.cgi?article=3539&context=etd (accessed on 27 March 2023).
36. Wu, C.; Zhang, Z.; Zheng, W. A Twisted and Coiled Polymer Artificial Muscles Driven Soft Crawling Robot Based on Enhanced Antagonistic Configuration. *Machines* **2022**, *10*, 142. [CrossRef]

37. Zhou, D.; Zuo, W.; Tang, X.; Deng, J.; Liu, Y. A multi-motion bionic soft hexapod robot driven by self-sensing controlled twisted artificial muscles. *Bioinspiration Biomim.* **2021**, *16*, 045003. [CrossRef]
38. Konda, R.; Bombara, D.; Swanbeck, S.; Zhang, J. Anthropomorphic Twisted String-Actuated Soft Robotic Gripper with Tendon-Based Stiffening. *IEEE Trans. Robot.* **2022**, *39*, 1178–1195. [CrossRef]
39. Sun, J.; Tighe, B.; Liu, Y.; Zhao, J. Twisted-and-Coiled Actuators with Free Strokes Enable Soft Robots with Programmable Motions. *Soft Robot.* **2021**, *8*, 213–225. [CrossRef] [PubMed]
40. Yip, M.C.; Niemeyer, G. High-performance robotic muscles from conductive nylon sewing thread. In Proceedings of the 2015 IEEE International Conference on Robotics and Automation (ICRA), Seattle, WA, USA, 26–30 May 2015; pp. 2313–2318. [CrossRef]
41. Luong, T.; Seo, S.; Hudoklin, J.; Koo, J.C.; Choi, H.R.; Moon, H. Variable Stiffness Robotic Hand Driven by Twisted-Coiled Polymer Actuators. *IEEE Robot. Autom. Lett.* **2022**, *7*, 3178–3185. [CrossRef]
42. Saharan, L.; De Andrade, M.J.; Saleem, W.; Baughman, R.H.; Tadesse, Y. iGrab: Hand orthosis powered by twisted and coiled polymer muscles. *Smart Mater. Struct.* **2017**, *26*, 105048. [CrossRef]
43. Tsabedze, T.; Hartman, E.; Zhang, J. A compact, compliant, and biomimetic robotic assistive glove driven by twisted string actuators. *Int. J. Intell. Robot. Appl.* **2021**, *5*, 381–394. [CrossRef]
44. Van der Weijde, J.; Smit, B.; Fritschi, M.; van de Kamp, C.; Vallery, H. Self-Sensing of Deflection, Force, and Temperature for Joule-Heated Twisted and Coiled Polymer Muscles via Electrical Impedance. *IEEE/ASME Trans. Mechatron.* **2017**, *22*, 1268–1275. [CrossRef]
45. Zhou, D.; Liu, Y.; Tang, X.; Zhao, J. Differential Sensing Method for Multidimensional Soft Angle Measurement Using Coiled Conductive Polymer Fiber. *IEEE Trans. Ind. Electron.* **2020**, *68*, 401–411. [CrossRef]
46. Wu, L.; Chauhan, I.; Tadesse, Y. A Novel Soft Actuator for the Musculoskeletal System. *Adv. Mater. Technol.* **2018**, *3*, 1700359. [CrossRef]
47. Zhang, P.; Li, G. Healing-on-demand composites based on polymer artificial muscle. *Polymer* **2015**, *64*, 29–38. [CrossRef]
48. Kim, S.H.; Lima, M.D.; Kozlov, M.E.; Haines, C.S.; Spinks, G.M.; Aziz, S.; Choi, C.; Sim, J.H.; Xuemin, W.; Hongbin, L.; et al. Harvesting temperature fluctuations as electrical energy using torsional and tensile polymer muscles. *Energy Environ. Sci.* **2015**, *8*, 3336–3344. [CrossRef]
49. Zarrouk, D.; Sharf, I.; Shoham, M. Analysis of worm-like robotic locomotion on compliant surfaces. *IEEE Trans. Biomed. Eng.* **2011**, *58*, 301–309. [CrossRef]

Article

Snakeskin-Inspired 3D Printable Soft Robot Composed of Multi-Modular Vacuum-Powered Actuators

Seonghyeon Lee [1,†], Insun Her [2,†], Woojun Jung [1] and Yongha Hwang [1,*]

1 Department of Control and Instrumentation Engineering, Korea University, Sejong 30019, Republic of Korea
2 Department of Electro-Mechanical Systems Engineering, Korea University, Sejong 30019, Republic of Korea
* Correspondence: hwangyongha@korea.ac.kr
† These authors contributed equally to this work.

Abstract: A modular soft actuator with snakeskin-inspired scales that generates an anisotropic friction force is designed and evaluated in this study. The actuator makes it possible to fabricate soft robots that can move on various surfaces in the natural environment. For existing modulus soft robots, additional connectors and several independent pneumatic pumps are required. However, we designed precise connection and snake-scale structures integrated with a single pneumatic modular actuator unit. The precise structure was printed using a DLP 3D printer. The movement characteristics of the soft robot changed according to the angle of the scale structure, and the movement distance increased as the number of modular soft actuator units increased. Soft robots that can move in operating environments such as flat land, tubes, inclined paths, and water have been realized. Furthermore, soft robots with modularization strategies can easily add modular units. We demonstrate the ability to deliver objects 2.5 times heavier than the full weight of the soft robot by adding tong-like structure to the soft robot. The development of a soft robot inspired by snakeskin suggests an easy approach to soft robots that enables various tasks even in environments where existing robots have limited activity.

Keywords: modular soft robot; snakeskin inspiration; keeled scales; anisotropic friction force; vacuum-powered actuators; DLP 3D printing

Citation: Lee, S.; Her, I.; Jung, W.; Hwang, Y. Snakeskin-Inspired 3D Printable Soft Robot Composed of Multi-Modular Vacuum-Powered Actuators. *Actuators* **2023**, *12*, 62. https://doi.org/10.3390/act12020062

Academic Editor: Hamed Rahimi Nohooji

Received: 9 January 2023
Revised: 27 January 2023
Accepted: 29 January 2023
Published: 31 January 2023

1. Introduction

Biological inspiration designs that mimic biological structures enable efficient movement in cluttered or unstructured environments such as medical care, search and rescue, and disaster response, as well as efficiency in tasks that require being in close contact with people [1,2]. Recently, soft robots have been constructed to mimic the structures of various mollusks [3]. In particular, a driving mechanism simulating the movement of snakes or caterpillars has shown effective movement in environments that impose motion constraints on wheeled or legged robots [4,5]. The caterpillar moves in one way using a controllable proleg and thoracic leg [6,7]. Snakes, despite having no limbs, can move by pulling forward scales of tile-like shapes that generate anisotropic friction and constrict the flexible body [8,9].

Various manufacturing methods, such as the reproduction molding process [10], rotary casting [11], screen printing [12], and 3D printing [13], have been proposed for manufacturing soft robots inspired by the living body [14,15]. In particular, owing to the development of 3D printing technology, the production of complex 3D structures by printing various polymer materials with high resolution has become possible [16]. Fused deposition modeling (FDM) technology, one of the most typical 3D printing methods, allows for the manufacture of a pneumatic actuator at a relatively low price using a thermoplastic elastomer (TPE) and thermoplastic polyurethane (TPU) material with a shore hardness of 70–80 A, which is more flexible than ordinary hard lactic acid (PLA) and acrylonitrile butadiene styrene (ABS) materials [17,18]. However, FDM 3D printers have limited resolution depending on the nozzle diameter (minimum: 0.1 mm); therefore, the thickness of the thin film constituting the actuator must be at least three times the size of the nozzle [19]. Direct ink writing (DIW) technology has been used to manufacture pneumatic actuators by extruding polymer materials such as silicone elastomers rather than filament-type materials [20,21]. However, similar to the FDM, the resolution is limited depending on the nozzle diameter and is inversely proportional to the build speed [19]. Polyjet technology is a technique for selectively spraying and depositing droplets of raw material onto the surface and has been used in the development of soft actuators using multi-material 3D printers that can print mixtures of rubber-like, hard, and soft materials [22,23]. However, as commercially manufactured inkjet materials are still limited, and designing chemicals capable of droplet formation and deposition and replacing printer heads is expensive, polyjet technology has drawbacks [19]. Stereolithography apparatus (SLA) technology selectively photopolymerizes and stacks liquid resins using a laser and can be used to develop micro-actuators because it can quickly and easily print high-resolution materials [24,25]. However, to date, multiple materials cannot be printed in a single printing sequence [26]. In addition, microsized soft actuators and fluid devices have been manufactured using casting polymer materials and chemically removing molds by printing molds with a 3D wax printer with a microsized resolution, rather than direct printing of polymer materials. However, mold removal technology using a 3D wax printer requires a relatively longer process time than a 3D printing method that directly prints polymer materials owing to the additional manufacturing process required to remove supporter materials and cure thermosetting polymers [27,28].

Soft pneumatic actuators can be divided into positive- and negative-pressure actuators. Positive-pressure soft actuators are inflated by injecting air-like fluid during operation, while negative-pressure soft actuators vacuum the interior and contract the volume of the actuator. Such negative-pressure actuators are preferred over positive-pressure actuators when performance limitations are required in terms of output force and deformation [29]. Compared to expanding positive-pressure actuators, negative-pressure actuators are reduced in volume during operation, making them unsuitable for use in narrow spaces and safe for use in vulnerable places such as the interior of people [30]. In addition, it is safer and less likely to fail than a positive-pressure soft actuator that expands during activation and is prone to breakage at high pressures [31].

The convenience of 3D structure manufacturing through 3D printing makes it easy to implement a modular strategy that disassembles one product into units that can be configured, decomposed, replaced, and managed easily [32]. This makes it easy to rebuild various forms to satisfy the requirements of a given environment and task. An assembly module system was used to implement soft robots such as gripping robot arms [33] and biomimicry moving robots [34] whose implementation is difficult with a single actuator. The reconfigurable module system can quickly replace units of various functions compared with single functions, enabling the implementation of a general-purpose soft robot. Using a reconfigurable module system, Zhang developed a soft robot capable of configuring worm-type, cross-legged quadrupedal walks, and hexapods [35]. Jiao developed robots with various functions such as gripping, bottle opening, crawling, and pipe climbing [36]. These modular soft robots adopt a method of applying pneumatic pressure to each of the connected modular actuator units, thus requiring several pneumatic pumps that are independently activated to perform the task [37,38]. On the other hand, Tawk developed a single pneumatic modular soft actuator to demonstrate the implementation of soft robots capable of the movement and rapid replacement of units with various functions. However, because parts made of other materials such as small rare-earth rings, rod magnets, and small plastic tubes are required to connect modular actuators, the fabrication process is complicated and limited to scale-down owing to manual assembly [39].

We have herein demonstrated a micro soft mobile robot that is as easy, simple, and free from breakage as possible. The soft robot is printed to a flexible resin using a high-resolution 3D printer using the digital light processing (DLP) method, which is a commercially available SLA method and is then completed after a typical postprocessing process. For simple fabrication, the robot was designed to have a shape that does not have a supporting structure, which is required for the implementation of complex structures in 3D printing technology. Through such 3D printing technology, precise connection structure can be produced, and a modularization strategy was applied to the robot. By repeatedly connecting basic modularized units of the same shape to form one robot that moves forward, composing the body of commercially available flexible resins is easy and simple. This is because even if a part of the body is defective during production, recreating the entire body is not necessary, and only one modular unit needs to be produced. In addition, a modular soft-moving robot capable of multiple connections can vary its length and mobility of the soft robot depending on the number of units connected by the user due to its modular structure. In particular, the modular soft actuator operates using a single pump connected to a single tube. As negative pressure is used, the actuator is free from the risk of bursting or leaking pressure compared with the method of operating by expansion due to positive pressure. To this end, it is designed to move only with linear contraction and restoration using a snakeskin structure application that generates an anisotropic friction force on the soft robot. The moving characteristics of a modularized snakeskin soft robot (MSSR) using a single pneumatic method have been experimentally confirmed according to design parameters, and its operation in practical operating environments, such as bent tubes and inclined paths, has been proven.

2. Concept and Design

2.1. Bioinspiration from the Snakeskin

The shape and size of scales on the skin of snakes vary depending on their type, habitat, and ecology. We designed MSSR scales inspired by the distinct keeled scales of *Atheris hispida*, also known as the Bush viper, which lives in rainforests, as shown in Figure 1a. Bush vipers, which can move in various environments (including trees, plant leaves, rocks, and soil) with anisotropic friction on surfaces of various roughness, have evolved efficiently; as a result, they have a distinct keeled appearance compared to other snake scales [40,41]. To realize such keeled scales, the 3D design of each scale was designed to ensure that the front surface in the moving direction would be slippery and the back surface in the opposite direction would be fixed. To make each of the designed scales converge at the center point of one of the pentagonal corners and form an anchor structure in the opposite direction of movement, we trimmed the back face into a triangular column. In such a design, a slippery smooth surface is formed on the front surface, a fixed anchor surface is formed on the rear surface, and the shape is similar to that of a snake scale, generating an anisotropic friction force on the floor.

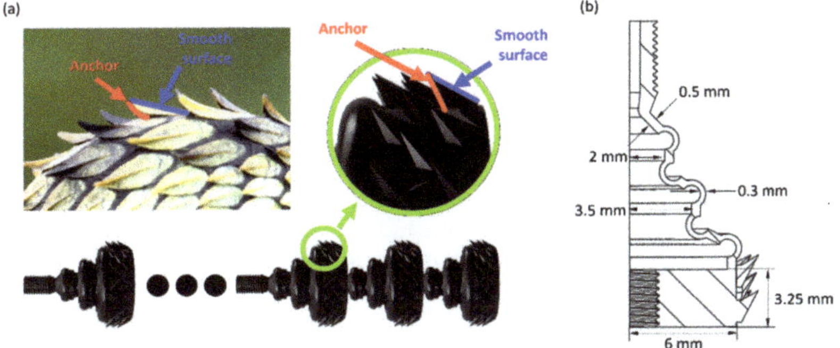

Figure 1. Schematic of MSSR. (**a**) MSSR's structure inspired by Bush viper's distinct keeled scales. (Photo courtesy of Mark Kostich, Adobe Stock images). (**b**) Cross-section (axial symmetry) and parameter values of soft actuator design for DLP 3D printing without supporting structures.

2.2. Structure Design and Fabrication of the MSSR

A flexible resin (F80, Dongguan Godsaid Technology, Dongguan, Guangdong Providence, China) was used to print the pneumatic soft actuator using a 3D printing method. The flexible resin has a shore hardness range of 50–60 A; thus, it exhibits high flexibility and resilience, and is suitable for making soft actuators. We empirically verified the printing performance of a recently commercialized DLP 3D printer (Sonic Mini 8 K, Phrozen Technology, Hsinchu City, Taiwan) with a 22-µm resolution and the properties of the tearable resin by creating a modular unit with as thin a wall as possible using 3D CAD (Fusion 360, Autodesk, Mill Valley, CA, USA). The thickness and dimensions of the unit are shown in Figure 1b. As support material different from the structural material cannot be used in a DLP printer for printing only a single material, a modular soft actuator unit was designed in a shape where the volume gradually decreased as the modular soft actuator unit goes upward and thus requires no internal supporting structure. A male connector and a female connector for assembling units were arranged on the top and bottom surfaces of the basic units so that multiple units can be easily connected. When we designed the connectors of the two parts, we empirically checked the diameters that can be connected without air leaking. Values of 0.125 inches for the male and 0.1325 inches for the female screw diameter were selected, and vacuum pressure was applied up to 15 kPa after sealing, but no problem was observed. An empty channel existed at the center of the male connector to apply air pressure to the actuator. Considering such conditions, we set the resin printing parameters: layer height (50 µm), exposure time (3 s), and rest time after retraction (4 s). After the completion of printing of each unit, each unit was removed from the platform of the 3D printer, and the stuck resin that was not hardened was removed by placing it in ethanol and stirring for 20 min. The unit was then placed in a UV curing device under a vacuum for 300 s for final curing. By connecting the circular symmetric modular actuator unit manufactured by 3D printing, the two design connectors of the integrated body engage through the thread so that air does not leak from the connection when air pressure is applied. Only the actuator unit placed at the front of the moving direction was designed without a female connector; thus, the free space of the MSSR was sealed. The back surface of the unit, arranged at the rear of the moving direction, was connected to a pressure pump to drive through a silicone tube.

2.3. Mechanism for Operation

The MSSR has a collapsible funnel design that folds vertically in the direction of travel, as shown in Figure 2a, to employ the negative-pressure system. A pneumatic soft actuator of the positive pressure type may be damaged by excessive expansion and may also damage the surrounding environment. To prevent such problems, a negative-pressure system that does not excessively expand was selected. The folding funnel structure used in the MSSR was designed to fold the wall without expanding or contracting; it was composed of flexible resin, so it could expand and contract in when operated under positive pressure. However, owing to the thin wall, the wall of MSSR may be torn or damaged if it expands under excessive pressure.

To effectively contract the actuator by folding the wall with vacuum pressure, a structure with an arc-shaped cross-section for each part where the wall surface was folded, as shown in Figure 2b, was added to the folding funnel structure. We compared the structure of the actuator with and without an arc joint through a finite element analysis using the solid mechanics modules of the multiphysics analysis program (COMSOL/Multiphysics, COMSOL, Burlington, MA, USA). First, a 3D actuator with a structure similar to that of an existing folding funnel was designed, and a structure containing an arc joint was added. The material properties were set at Young's Modulus 2.159 MPa in consideration of Shore-A hardness of F80 resin [42]. In addition, tensile strength (3.5 MPa), tear strength (9.7 kN/m), elongation at break (159%), and viscosity (2300 mPa·s) were used from the datasheet of F80. A Poisson ratio of material was set by referring to a general flexible resin physical property of 0.49 [43]. The density was measured from the volume and mass of the printed resin structure. By applying the same negative pressure inside each actuator, the changes in the two 3D structures and the displacement of one point on the top surface were analyzed; subsequently, the 3D printed actuator was air pressure tested and measured. By comparing the results of the pneumatic test of actuators manufactured in the two designs in Figure 2c with pressure-induced displacement change, we observed that actuators with arc joints showed 1041.6% higher displacement than actuators without arc joints and had a 9.3% fine error with simulation analysis. This is because the arc joints facilitate the folding of the actuator wall and reduce the elastic force generated by the folding.

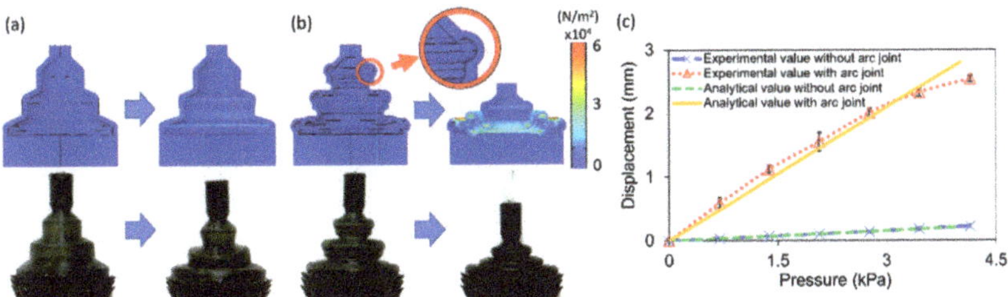

Figure 2. FEM analysis results and the actuator with vacuum applied for characterization of the soft actuator: (**a**) Collapsible funnel-shaped design. (**b**) Arc joint design. (**c**) Displacement of actuators depending on applied pressure.

Figure 3a illustrates the progression mechanism of MSSR. In the process of contracting or returning to the initial state, when five soft actuator units were connected, the snake scales of each unit generated different frictional forces from the floor. When contracting initially, negative pressure was applied to the MSSR, and the scales of the rear unit of the moving direction based on the center of gravity demonstrated a phenomenon wherein a smooth surface with low friction force brushed the floor and slipped. In the front unit, on the contrary, the anchor part of the scales was fixed by generating high friction force with the floor. Accordingly, the front unit showed fixed or low movement, and the rear unit slid and exhibited high mobility. If the contracted amount of air was placed inside again, it returned to its original state; however, in contrast to the contraction, the rear unit was fixed, and the front unit slid, indicating a phenomenon wherein the whole body moved forward. By repeating one series of cycles, the MSSR to which multiple units were connected could proceed as intended. After connecting a micro silicone hose (inside diameter: 0.4 mm; outside diameter: 1.2 mm) to the manufactured unit, such as Figure 3b, an air pressure test with a syringe confirmed that it moved in the same manner (Video S1).

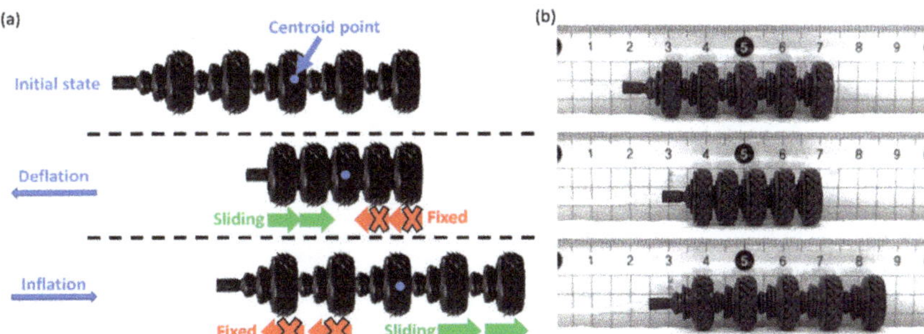

Figure 3. Locomotion model of MSSR. (**a**) Movement mechanism of MSSR with five modular soft actuator units connected. (**b**) Images of the moving MSSR during one cycle.

3. Discussion

We considered the effect of the angle formed between the rear anchor part of the snake scales and horizontal axis, as shown in Figure 4. As shown in Figure 4a, actuator units with scales between the horizontal axes and anchor parts ranging from 30° to 60° were manufactured, and their moving capabilities on wood and paper substrates were compared with MSSRs with five connected units. The term "iteration" denotes the number of cycles that completely contract and return to the original state. As seen in Figure 4b, the MSSR on the paper substrate MSSRs with 30° scales demonstrated the highest displacement of the four designs. On the wood substrate, the MSSR at the 50° scale demonstrated the highest displacement, as shown in Figure 4c. Paper substrates with lower roughness than wooden plates enabled us to confirm the characteristics of the MSSR movement on a smooth surface. The changes in movement characteristics of MSSR on 400 grit and 100 grit sandpaper with 13.21 μm and 24.93 μm roughness are shown in Figure 4d,e, respectively. In addition, to confirm the anisotropic friction force of each scale generated on the surface of the paper, wooden, and sandpaper boards, the static friction coefficient was determined, as shown in Table 1. It shows the difference between the forward friction coefficient ($\mu_{S.forward}$) and the backward friction coefficient ($\mu_{S.backward}$) according to the scale angle and the substrate material. The differences in the friction coefficient and displacement of the MSSR for each substrate showed the same tendency. Therefore, the characteristics of the displacement varied depending on the roughness of the floor surface when the MSSR moved. As the anchor surface is hardly fixed on the smooth surface, the unit needs to have fixed slips. Owing to these characteristics, replacing actuator units with appropriate snake-scale angles considering the roughness of the environment in which the MSSR operates is appropriate. Module replacement depending on the situation is simple and easy owing to the module strategy.

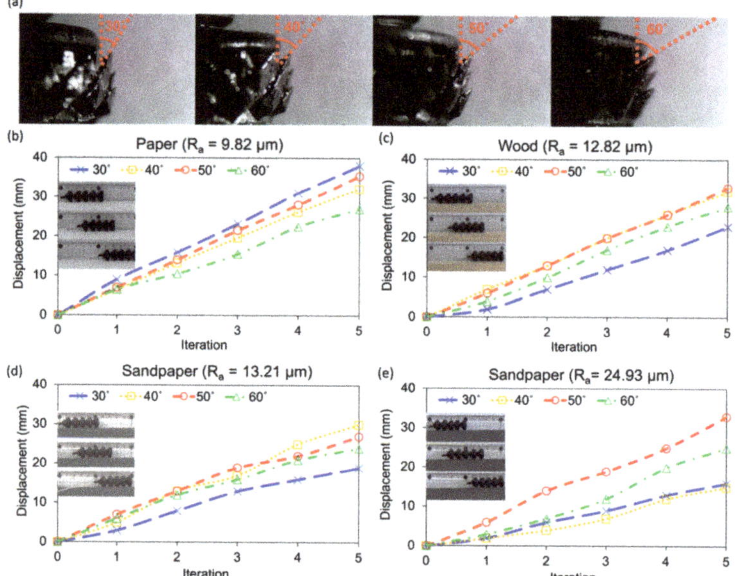

Figure 4. Characteristics depending on the angle between the horizontal axis and the anchor plane of the snake scale structure. (**a**) Close-up image of snake scale structure designed from 30° to 60° (red dotted line representing the angle of the anchor face of the calculated 3D design). Displacement according to the operating cycle of MSSR, which is made of five modular soft actuator units, on (**b**) paper, (**c**) wood surfaces and sandpapers with two different surface roughness ((**d**) 13.21 μm, (**e**) 24.93 μm).

Table 1. Differences between backward and forward friction coefficients ($\mu_{S.forward} - \mu_{S.backward}$) according to snake scale structure angle.

Angle	30°	40°	50°	60°
Wood (R_a = 12.82)	0.14	0.28	0.30	0.20
Paper (R_a = 9.82)	0.23	0.06	0.17	0.03
Sandpaper (R_a = 13.21)	0.25	0.59	0.48	0.38
Sandpaper (R_a = 24.93)	0.18	0.15	1.02	0.67

As the number of modular actuator units increased, the displacement per iteration generated in the process of contraction or returning to the initial state increased, as shown in Figure 5. This is because the total displacement increases in proportion to the number of units that are resistant when pushed backward but easily slip forward. However, in a slippery material such as glass, where an anisotropic frictional force due to the scale structure is not generated, the change in displacement does not increase, even when the number of units increases. The experiment was conducted on a paper towel with a rough surface prone to anisotropic frictional forces.

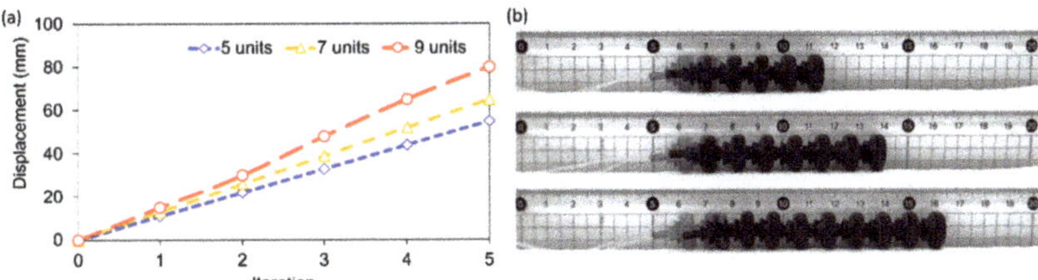

Figure 5. Characteristics according to the number of modular soft actuator units. (**a**) Displacement changes according to the operation cycle of MSSR with different number of units on the rough surface paper towel that easily has anisotropic frictional force. (**b**) Initial and final positions of the captured MSSR after one cycle.

The advantage of modular robots is that they can replace units with various functions. In particular, to put mobile soft robots into practical use, heavy parts and cargo can be transported compared with the main body. Therefore, we added units in the shape of tongs at the front of the MSSR that are designed to fix a payload that moves with the unit, as shown in Figure 6 and Video S2. As the number of connected units increased, the weight of the payload that could be moved without MSSR bending also increased. We demonstrated that if there are 11 units, up to 2.5 times the weight of the MSSR can be moved as a payload. This is because as each unit increases, the total friction force generated between the anchor surface of the scale and the bottom increases compared with the gravity caused by the weight and the actuator itself. If the MSSR exceeds the weight of the payload it can push, it will not move forward as it will bend its entire body, or all units will slip.

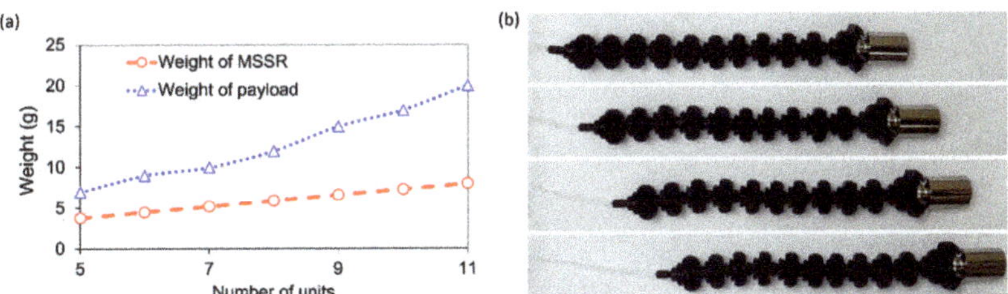

Figure 6. MSSR capable of delivering payload. (**a**) Change in payload weight that can be moved according to the number of units by adding a tong-type unit. (**b**) Images captured according to the cycle of delivering a weight 2.5 times the weight of the entire MSSR.

Soft robots that can move without arms and legs are suitable for operation in narrow environments such as pipelines. Therefore, the MSSR has a cylindrical design that can be moved even in a tubular space. We demonstrated MSSR movement with soft transparent PVC tubes, as shown in Figure 7 and Video S3. In addition, Figure 7a shows that it could successfully pass through a bent tube. This demonstrates the strength of the soft actuator, allowing the MSSR unit to operate in a bent state by being made of flexible resin. Figure 7b shows that the MSSR can not only move smoothly in curved tubes but also in S-shaped tubes. The technology of soft robots in structures inspired by snake skins shows the practicality of advanced-level operations in living bodies through 3D printer technology, and supports high-resolution biocompatible material printing in the future.

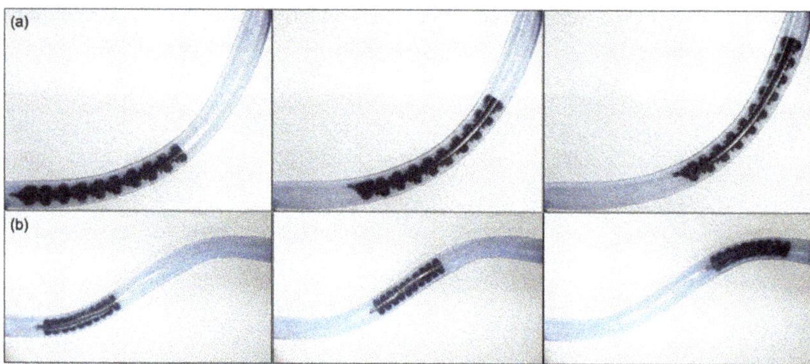

Figure 7. Demonstration of MSSR in a transparent PVC tube. Images captured according to the period of images captured in the (**a**) bent tube and (**b**) S-curved tube.

Soft robots should not only be able to operate in planes or tubes but also in a variety of environments. In particular, the natural environment has hill-like terrain with slopes, which can limit movement. Accordingly, the movement of the soft robot on a substrate with different slopes, as shown in Figure 8a, was demonstrated. However, on a substrate where sufficient friction force is not generated, the entire soft robot slips according to the inclination, and the soft robot cannot move forward. In the slope of the paper towel surface, MSSR showed a maximum angle of 30°. At the slope angle of 30°, the moving distance of one cycle of the MSSR was reduced by 33.33% compared to the untilted plane because sliding due to gravity in the slope tends to reduce the moving distance. In addition, in the natural environment, pools of water may occur on flatlands depending on weather or ecology. Such humid environments can cause fatal failures in electrically operated robots. The MSSR is suitable for underwater operation because leakage of internal and external fluid does not occur through the engagement of the thread of the 3D printed connection. Figure 8b demonstrate the mobility of the MSSR underwater. An MSSR, whose interior is filled with air, floats on the surface by buoyancy, but by pressing water pressure with a syringe when the interior is filled with water, anisotropic movement is shown in the same manner as in a flat surface. The speed of movement in the water decreased by 55.88% compared to that of paper towels because there is a drag force of water. Mobility in such environments widens the scope of the environment in which modular units with various functions can perform tasks.

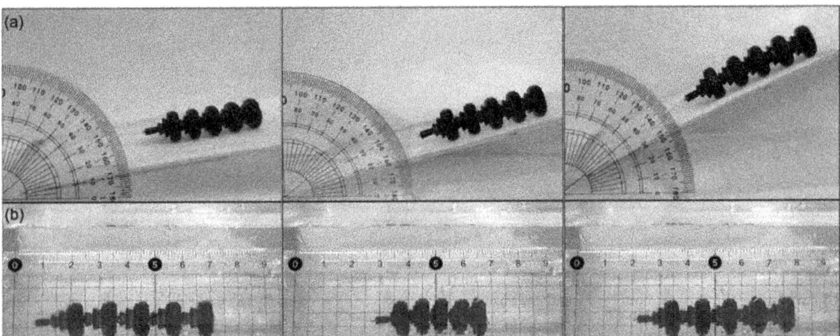

Figure 8. Locomotion of MSSR under different conditions. (**a**) Demonstration of movement along the slope of the paper tower surface. (**b**) Captured pictures according to the operating cycle of the MSSR under water.

4. Conclusions

In this study, a modular soft robot was designed and manufactured based on snake scales to enable movement using a single pneumatic actuator. The transfer characteristics of the MSSR by the structure imitating the keeled scales that generate anisotropic friction in various environments were confirmed through changes in the angle with horizontal lines and the number of units of scales. In addition, a tong-type unit was added to confirm the advantage of the modular robot operating by adding modular units with various functions. In the case of adding the tong structure, when the number of units is 11, it can deliver more than 2.5 times the weight of the main body of the MSSR. The MSSR, which is made of flexible resin, demonstrated the advantages of flexible robots by demonstrating bending and moving in cylindrical, curved, and S-shaped tubes. In addition, demonstrating the movement of the robot on slopes and in water showed the possibility of mobility in various environments. In future research, pneumatic actuators will be added to both sides of the module in the head of the robot to allow a smooth change of direction of the MSSR. This diversity of operating environments demonstrates the possibility of developing a soft robot that not only allows robots to move in difficult environments where robots with arms and legs are restricted but also allows for search, rescue, exploration, and inspection work through simple unit replacement. The module was repeatedly tested 10,000 times at constant pressure to confirm that there was no reliability issue. In addition, it was confirmed that the displacement decreased by 3.34% on average at the same pressure using MSSR samples that were left at room temperature for 7 days and 80 days, respectively. This shows that there is no drawback due to material properties. Subsequently, with the development of 3D printing, we believe that microsize implementation and chemically and biologically stable materials can be utilized within the scope of medical and biological fields.

Supplementary Materials: The following supporting information can be downloaded at: https://www.mdpi.com/article/10.3390/act12020062/s1; Video S1: Operation of the MSSR with five modular soft actuator units connected; Video S2: Demonstrated delivery of a payload 2.5 times heavier than the full MSSR weight; Video S3: Repeated operation of MSSR in the curved tube and S-curved tube.

Author Contributions: Conceptualization, S.L., I.H., W.J. and Y.H.; methodology, S.L., W.J. and Y.H.; validation, S.L., I.H. and Y.H.; writing—original draft preparation, S.L. and I.H.; writing—review and editing, Y.H.; visualization, I.H. and W.J.; supervision, Y.H.; project administration, Y.H. All authors have read and agreed to the published version of the manuscript.

Funding: This research was supported by the National Research Foundation of Korea (NRF), funded by the Korean Government (MIST) under Grant NRF-2022R1F1A1074083, and the "Regional Innovation Strategy (RIS)" through the NRF, funded by the Ministry of Education under Grant 2021RIS-004.

Institutional Review Board Statement: Not applicable.

Informed Consent Statement: Not applicable.

Data Availability Statement: The data presented in this study are available on request from the corresponding author.

Conflicts of Interest: The authors declare no conflict of interest.

References

1. Trivedi, D.; Rahn, C.D.; Kier, W.M.; Walker, I.D. Soft robotics: Biological inspiration, state of the art, and future research. *Appl. Bion. Biomech.* **2008**, *5*, 99–117. [CrossRef]
2. Kim, S.; Laschi, C.; Trimmer, B. Soft robotics: A bioinspired evolution in robotics. *Trends Biotechnol.* **2013**, *31*, 287–294. [CrossRef] [PubMed]
3. Nurzaman, S.G.; Iida, F.; Margheri, L.; Laschi, C. Soft robotics on the move: Scientific networks, activities, and future challenges. *Soft Robot.* **2014**, *1*, 154–158. [CrossRef]
4. Wang, C.; Puranam, V.R.; Misra, S.; Venkiteswaran, V.K. A Snake-Inspired Multi-Segmented Magnetic Soft Robot Towards Medical Applications. *IEEE Robot. Autom. Lett.* **2022**, *7*, 5795–5802. [CrossRef]
5. Zhang, B.; Fan, Y.; Yang, P.; Cao, T.; Liao, H. Worm-Like Soft Robot for Complicated Tubular Environments. *Soft Robot.* **2019**, *6*, 399–413. [CrossRef]
6. Trimmer, B.; Issberner, J. Kinematics of soft-bodied, legged locomotion in Manduca sexta larvae. *Biol. Bull.* **2007**, *212*, 130–142. [CrossRef]
7. Niu, H.; Feng, R.; Xie, Y.; Jiang, B.; Sheng, Y.; Yu, Y.; Baoyin, H.; Zeng, X. MagWorm: A Biomimetic Magnet Embedded Worm-Like Soft Robot. *Soft Robot.* **2021**, *8*, 507–518. [CrossRef]
8. Hu, D.L.; Nirody, J.; Scott, T.; Shelley, M.J. The mechanics of slithering locomotion. *Proc. Natl. Acad. Sci. USA* **2009**, *106*, 10081–10085. [CrossRef]
9. Marvi, H.; Cook, J.P.; Streator, J.L.; Hu, D.L. Snakes move their scales to increase friction. *Biotribology* **2016**, *5*, 52–60. [CrossRef]
10. Shepherd, R.F.; Ilievski, F.; Choi, W.; Morin, S.A.; Stokes, A.A.; Mazzeo, A.D.; Chen, X.; Wang, M.; Whitesides, G.M. Multigait soft robot. *Proc. Natl. Acad. Sci. USA* **2011**, *108*, 20400–20403. [CrossRef]
11. Polygerinos, P.; Lyne, S.; Wang, Z.; Nicolini, L.F.; Mosadegh, B.; Whitesides, G.M.; Walsh, C.J. Towards a soft pneumatic glove for hand rehabilitation. In Proceedings of the 2013 IEEE/RSJ International Conference on Intelligent Robots and Systems, Tokyo, Japan, 3–7 November 2013; pp. 1512–1517.
12. Yeo, J.C.; Yap, H.K.; Xi, W.; Wang, Z.; Yeow, C.-H.; Lim, C.T. Flexible and Stretchable Strain Sensing Actuator for Wearable Soft Robotic Applications. *Adv. Mater. Technol.* **2016**, *1*, 1600018. [CrossRef]
13. Gul, J.Z.; Sajid, M.; Rehman, M.M.; Siddiqui, G.U.; Shah, I.; Kim, K.-H.; Lee, J.-W.; Choi, K.H. 3D printing for soft robotics—A review. *Sci. Technol. Adv. Mater.* **2018**, *19*, 243–262. [CrossRef]
14. Schmitt, F.; Piccin, O.; Barbé, L.; Bayle, B. Soft Robots Manufacturing: A Review. *Front. Robot. AI* **2018**, *5*, 84. [CrossRef]
15. Dong, X.; Luo, X.; Zhao, H.; Qiao, C.; Li, J.; Yi, J.; Yang, L.; Oropeza, F.J.; Hu, T.S.; Xu, Q.; et al. Recent advances in biomimetic soft robotics: Fabrication approaches, driven strategies and applications. *Soft Matter* **2022**, *18*, 7699–7734. [CrossRef]
16. Miller, J.S.; Stevens, K.R.; Yang, M.T.; Baker, B.M.; Nguyen, D.-H.T.; Cohen, D.M.; Toro, E.; Chen, A.A.; Galie, P.A.; Yu, X.; et al. Rapid casting of patterned vascular networks for perfusable engineered three-dimensional tissues. *Nat. Mater.* **2012**, *11*, 768–774. [CrossRef]
17. Yap, H.K.; Ng, H.Y.; Yeow, R.C.-H. High-Force Soft Printable Pneumatics for Soft Robotic Applications. *Soft Robot.* **2016**, *3*, 144–158. [CrossRef]
18. Hu, W.; Alici, G. Bioinspired Three-Dimensional-Printed Helical Soft Pneumatic Actuators and Their Characterization. *Soft Robot.* **2020**, *7*, 267–282. [CrossRef]
19. Wallin, T.J.; Pikul, J.; Shepherd, R.F. 3D printing of soft robotic systems. *Nat. Rev. Mater.* **2018**, *3*, 84–100. [CrossRef]
20. Plott, J.; Shih, A. The extrusion-based additive manufacturing of moisture-cured silicone elastomer with minimal void for pneumatic actuators. *Addit. Manuf.* **2017**, *17*, 1–14. [CrossRef]
21. Schaffner, M.; Faber, J.A.; Pianegonda, L.; Rühs, P.A.; Coulter, F.; Studart, A.R. 3D printing of robotic soft actuators with programmable bioinspired architectures. *Nat. Comm.* **2018**, *9*, 878. [CrossRef]
22. Kalisky, T.; Wang, Y.; Shih, B.; Drotman, D.; Jadhav, S.; Aronoff-Spencer, E.; Tolley, M.T. Differential pressure control of 3D printed soft fluidic actuators. In Proceedings of the 2017 IEEE/RSJ International Conference on Intelligent Robots and Systems (IROS), Vancouver, BC, Canada, 24–28 September 2017; pp. 6207–6213.
23. MacCurdy, R.; Katzschmann, R.; Youbin, K.; Rus, D. Printable hydraulics: A method for fabricating robots by 3D co-printing solids and liquids. In Proceedings of the 2016 IEEE International Conference on Robotics and Automation (ICRA), Stockholm, Sweden, 16–21 May 2016; pp. 3878–3885.

24. Peele, B.N.; Wallin, T.J.; Zhao, H.; Shepherd, R.F. 3D printing antagonistic systems of artificial muscle using projection stereolithography. *Bioinspir. Biomim.* **2015**, *10*, 055003. [CrossRef] [PubMed]
25. Kang, H.-W.; Lee, I.H.; Cho, D.-W. Development of a micro-bellows actuator using micro-stereolithography technology. *Microelectron. Eng.* **2006**, *83*, 1201–1204. [CrossRef]
26. Truby, R.L.; Lewis, J.A. Printing soft matter in three dimensions. *Nature* **2016**, *540*, 371–378. [CrossRef] [PubMed]
27. Lee, S.; Jung, W.; Ko, K.; Hwang, Y. Wireless Micro Soft Actuator without Payloads Using 3D Helical Coils. *Micromachines* **2022**, *13*, 799. [CrossRef] [PubMed]
28. Jung, W.; Lee, S.; Hwang, Y. Truly 3D microfluidic heating system with iterative structure of coil heaters and fluidic channels. *Smart Mater. Struct.* **2022**, *31*, 035016. [CrossRef]
29. Tawk, C.; Alici, G. A Review of 3D-Printable Soft Pneumatic Actuators and Sensors: Research Challenges and Opportunities. *Adv. Intell. Syst.* **2021**, *3*, 2000223. [CrossRef]
30. Yang, D.; Verma, M.S.; So, J.-H.; Mosadegh, B.; Keplinger, C.; Lee, B.; Khashai, F.; Lossner, E.; Suo, Z.; Whitesides, G.M. Buckling Pneumatic Linear Actuators Inspired by Muscle. *Adv. Mater. Technol.* **2016**, *1*, 1600055. [CrossRef]
31. Yang, D.; Verma, M.S.; Lossner, E.; Stothers, D.; Whitesides, G.M. Negative-Pressure Soft Linear Actuator with a Mechanical Advantage. *Adv. Mater. Technol.* **2017**, *2*, 1600164. [CrossRef]
32. Zhang, C.; Zhu, P.; Lin, Y.; Jiao, Z.; Zou, J. Modular Soft Robotics: Modular Units, Connection Mechanisms, and Applications. *Adv. Intell. Syst.* **2020**, *2*, 1900166. [CrossRef]
33. Phillips, B.T.; Becker, K.P.; Kurumaya, S.; Galloway, K.C.; Whittredge, G.; Vogt, D.M.; Teeple, C.B.; Rosen, M.H.; Pieribone, V.A.; Gruber, D.F. A dexterous, glove-based teleoperable low-power soft robotic arm for delicate deep-sea biological exploration. *Sci. Rep.* **2018**, *8*, 14779. [CrossRef]
34. Zhou, Y.; Jin, H.; Liu, C.; Dong, E.; Xu, M.; Yang, J. A novel biomimetic jellyfish robot based on a soft and smart modular structure (SMS). In Proceedings of the 2016 IEEE International Conference on Robotics and Biomimetics (ROBIO), Qingdao, China, 3–7 December 2016; pp. 708–713.
35. Zhang, Y.; Zheng, T.; Fan, J.; Li, G.; Zhu, Y.; Zhao, J. Nonlinear Modeling and Docking Tests of a Soft Modular Robot. *IEEE Access* **2019**, *7*, 11328–11337. [CrossRef]
36. Jiao, Z.; Zhang, C.; Wang, W.; Pan, M.; Yang, H.; Zou, J. Advanced Artificial Muscle for Flexible Material-Based Reconfigurable Soft Robots. *Adv. Sci.* **2019**, *6*, 1901371. [CrossRef]
37. Zou, J.; Lin, Y.; Ji, C.; Yang, H. A reconfigurable omnidirectional soft robot based on caterpillar locomotion. *Soft Robot.* **2018**, *5*, 164–174. [CrossRef]
38. Meng, C.; Xu, W.; Li, H.; Zhang, H.; Xu, D. A new design of cellular soft continuum manipulator based on beehive-inspired modular structure. *Int. J. Adv. Robot. Syst.* **2017**, *14*, 1729881417707380. [CrossRef]
39. Tawk, C.; Panhuis, M.I.H.; Spinks, G.M.; Alici, G. Bioinspired 3d printable soft vacuum actuators for locomotion robots, grippers and artificial muscles. *Soft Robot.* **2018**, *5*, 685–694. [CrossRef]
40. Ashe, J. A New Bush Viper. *J. East Afr. Nat. Hist.* **1968**, *1968*, 53–59. [CrossRef]
41. Groen, J.; Bok, B.; Tiemann, L.; Verspui, G.J. Geographic range extension of the rough-scaled bush viper, Atheris hispida (Serpentes: Viperidae) from Uganda, Africa. *Herpetol. Notes* **2019**, *12*, 241–243.
42. Larson, K. *Can You Estimate Modulus from Durometer Hardness for Silicones*; Dow Corning Corporation: Midland, MI, USA, 2016; pp. 1–6.
43. Chung, S.M.; Yap, A.U.J.; Koh, W.K.; Tsai, K.T.; Lim, C.T. Measurement of Poisson's ratio of dental composite restorative materials. *Biomaterials* **2004**, *25*, 2455–2460. [CrossRef]

 actuators

Article

Handshake Feedback in a Haptic Glove Using Pouch Actuators

Seiya Yamaguchi [1], Takefumi Hiraki [2], Hiroki Ishizuka [3] and Norihisa Miki [1,*]

[1] Department of Mechanical Engineering, Keio University, Yokohama 223-8522, Kanagawa, Japan
[2] Faculty of Library, Information and Media Science, University of Tsukuba, Tsukuba 305-8550, Ibaraki, Japan
[3] Graduate School of Engineering Science, Osaka University, Toyonaka 560-8531, Osaka, Japan
* Correspondence: miki@mech.keio.ac.jp

Abstract: In this paper, we propose and demonstrate a haptic device with liquid-pouch motors that can simulate a handshake. Because handshakes involve contact of the palms or soft skin, handshake simulation requires the haptic device to provide pressure onto specific areas of the palm with soft contact. This can be achieved with thermally driven liquid-pouch motors, which inflate and deflate when a low-boiling-point liquid, here Novec™ 7000, evaporates and condenses, respectively. Due to the simplicity of the soft actuator system, this haptic glove is lightweight and conformable. To design the haptic glove, we experimentally investigated the contact region and strength in handshakes, which led to an optimal number, size and position for the liquid-pouch motors. Sensory experiments with human subjects verified that the designed haptic glove successfully simulated handshakes.

Keywords: soft actuator; handshake; haptic glove; pouch actuator; social touch

Citation: Yamaguchi, S.; Hiraki, T.; Ishizuka, H.; Miki, N. Handshake Feedback in a Haptic Glove Using Pouch Actuators. *Actuators* **2023**, *12*, 51. https://doi.org/10.3390/act12020051

Academic Editor: Hamed Rahimi Nohooji

Received: 25 December 2022
Revised: 16 January 2023
Accepted: 19 January 2023
Published: 24 January 2023

1. Introduction

The skin is our largest organ and covers our whole body, and we particularly rely on tactile information obtained from the hands [1]. As shown in the cortical homunculus, a large area of the brain is used to process information from the hands [2]. Such information is used in the characterization and manipulation of touched objects and to interact and communicate with others.

Social touching, such as handshakes, stroking and hugs, is known to have positive effects on us in many aspects, such as physical and emotional, attachment, attitude and behavior and emotional transference [3]. In addition, social touch increases intimacy in interpersonal relationships because it can leave the impression that the other person is "more familiar" or "happy to see you" [4]. Devices that reproduce stroking or hugging motions have been demonstrated [5,6]. In this study, we focused on handshakes. Because tactile communication takes place only at the hands in a handshake, we considered replication of handshakes with a rather simple system, specifically, haptic gloves.

Haptic gloves have been developed as wearable devices to present tactile sensations to users' hands [7], which can enrich the user experience in extended-reality technology and human–computer interactions [8,9]. Haptic gloves stimulate skin or tactile receptors inside the skin either physically or electrically. For physical-type haptic gloves, they need to be compact and light enough to be worn by the users, while generating a sufficiently large force and displacement with the actuators of the gloves. Soft actuators are considered to contribute to such glove applications. A haptic glove using soft doughnut-shaped actuators was proposed to provide tactile feedback to the fingers [10]. A soft actuator that is composed of an anisotropically fiber-reinforced material and an elastomer matrix is attached to the back of the finger to provide force sensation to the user [11]. The haptic glove with this actuator can generate large feedback without interfering with the user's motion on the palm side. Pulse width modulation (PWM) of a small soft actuator embedded inside the glove was reported to efficiently provide tactile feedback [12]. Soft actuators can easily work with other actuators due to their lightweight. A pneumatic balloon actuator is used for force sensation in addition to an electromagnetic actuator that generates vibration to provide a greater sense of touch [13]. Thus, glove-type devices that present tactile sensations are expected. However, because the actuators are close to human skin, high voltage and temperature must be avoided and, in addition, soft or compliant actuators are preferable.

As soft and compliant actuators, pneumatic actuators have been studied. Pneumatic actuators are promising soft actuators [14,15]. Air is the working fluid, and it has lightweight and large compliance. The drawbacks of pneumatic actuators are that the air needs to be supplied externally using tubes and valves, which are typically bulky, from a control system.

Gas–liquid phase change actuators have been studied as compact soft actuators [16–22]. A low-boiling-point liquid, such as Novec ™ 7000 with a boiling point of 34 °C, is encapsulated inside a deformable pouch. When the pouch is heated, the encapsulated liquid vaporizes, and the pouch volume drastically increases. Centimeter-sized pouch motors were reported to produce a force of 50 N [21]. The pouch can be miniaturized down to the millimeter scale [23]. These liquid-pouch motors require only a heat source in order to be actuated. An external heat source, such as a hair dryer, does not require any additional components to control the pouch. Direct heating of the pouch by designing a resistive heater on or in the vicinity of the pouch is favored due to its high efficiency and ease of control [24]. When metal particles are included in the encapsulated liquid, it can be heated by irradiation using an infrared laser, which can control the pouch motor externally with high efficiency [25–27].

In this paper, we propose and demonstrate a haptic device with liquid-pouch motors containing Novec ™ 7000 to simulate a handshake. Figure 1 shows the concept figure of the haptic glove that simulates a handshake. Because handshakes involve contact of the palms or soft skin, the simulation of a handshake requires the haptic device to provide pressure onto specific areas of the palm with soft contact. This can be achieved with liquid-pouch motors. In addition, warm liquid-pouch motors, which are slightly warmer than the boiling point of Novec ™ 7000, may contribute by simulating the warmness of another person's hand. External heating with a hair dryer was used to actuate the liquid-pouch motors. We experimentally investigated the contact region and strength during handshakes, which were used to determine the number, size and position of the liquid-pouch motors. A questionnaire-based survey was conducted to validate the concept. This work was approved by the Research Ethics Committee of the Faculty of Science and Technology, Keio University (2021-89).

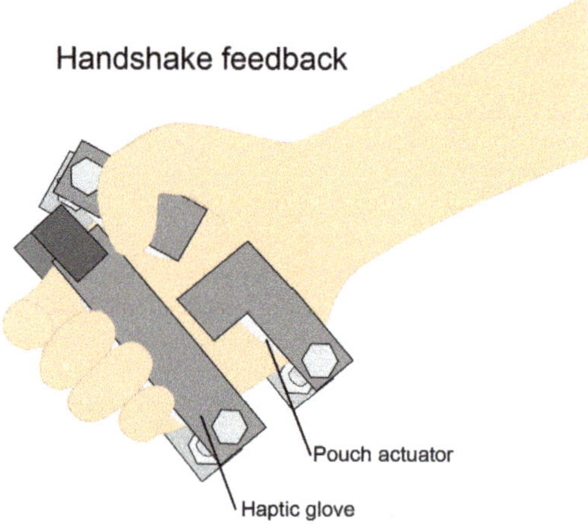

Figure 1. The haptic glove that simulates a handshake. The pouch actuators provide pressure onto the palm and the back of the hand to replicate the tactile perception in a handshake.

2. Design of Pouch Gloves

2.1. Pouch Actuator

The pouch actuator consists of a hydrofluoroether fluid (Novec ™ 7000, 3M Company, St. Paul, MN, USA) with a boiling point of 34 °C enclosed in a 75 µm-thick nylon polyethylene sheet (Nylon Poly New L Type, Fukusuke Kogyo, Niihama, Japan). To fabricate the product, we first cut two polyethylene sheets with scissors and heat-sealed three sides using a heat sealer (Ishizaki Electric Works, Tokyo, Japan). The width of the seal was 2.5 mm. We then injected a fixed amount of Novec ™ 7000 with a micro syringe and heat-sealed the remaining side to form a pouch. Because polyethylene does not expand or contract, the area to which force is applied can be controlled by the area of the pouch when enough Novec ™ 7000 is sealed within it.

The pouch actuator is used as a bistable thermal actuator; when heated to a temperature above the boiling point of Novec ™ 7000, the actuator expands due to the gas–liquid phase change of Novec ™ 7000. The pouch actuator is positioned on the palm of the user's hand with a haptic glove and when the actuator is heated, it expands and provides pressure on the palm. The temperature of the pouch actuator during expansion is slightly higher than 34 °C. Because the average surface temperature of the human body is about 33 °C, the inflated pouch roughly replicates the warmth of another person's hand. The response time of the pouch is 3~5 s after heating, and there is no problem with handshake feedback.

The size of the pouch actuator needs to be determined. To avoid disturbing the opening and closing of the wearer's hand, the length of the horizontal side of the pouch was set to be 20.0 mm. The length of the vertical side was from 20.0 to 50.0 mm. Based on the results of our preliminary experiments, when the amount of Novec ™ 7000 sealed in the pouch was 0.30 mL, the liquid phase of Novec ™ 7000 remained in the pouch even when the pouch was maximally inflated.

2.2. Overall Design of Haptic Glove

Figure 2 shows a photograph of the proposed device. As shown in Figure 3, the device is composed of a pouch actuator, a support base, and an adjustment screw. The pouch expands upon heating to provide a tactile pressure sensation to the palm.

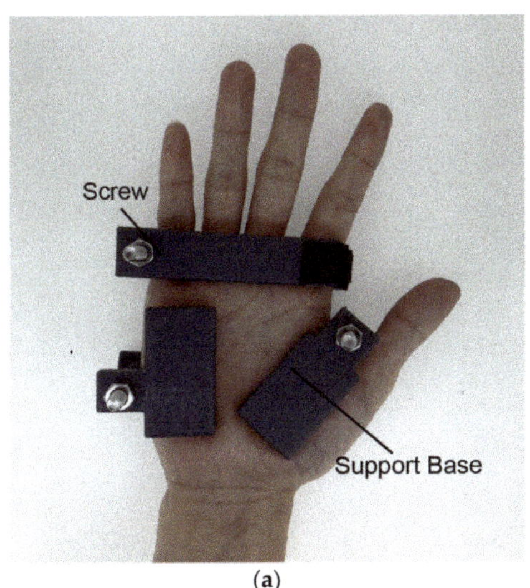

(a)

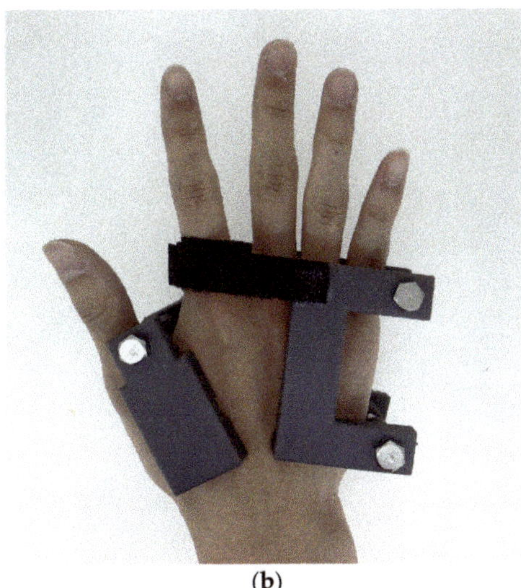

(b)

Figure 2. Photograph of proposed haptic glove: (**a**) palm side and (**b**) back side of the hand.

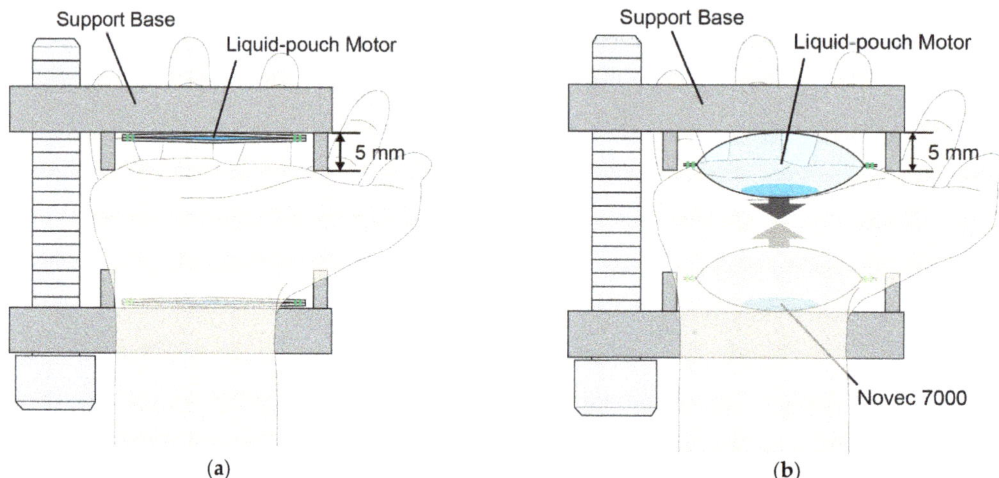

Figure 3. Mechanism of applying pressure onto palm with pouch actuators: (**a**) before heating and (**b**) after heating.

The support base has two plates with pockets inside for placing the pouch actuators, as shown in Figure 4. We designed the gap between the plates to be controlled with screws. We fabricated the plates with gray resin (UV-curable acrylic resin) using a 3D printer (Form3, Formlabs, Medford, OR, USA). The tensile strength and Young's modulus of the gray resin are 65 MPa and 2.8 GPa, respectively. The weight of this material is approximately 15 g, which is sufficiently light to avoid interference with hand movement. The resin does not deform when the pouch actuator expands and can hold the actuator in place. A video of the haptic glove can be found in Supplementary Materials.

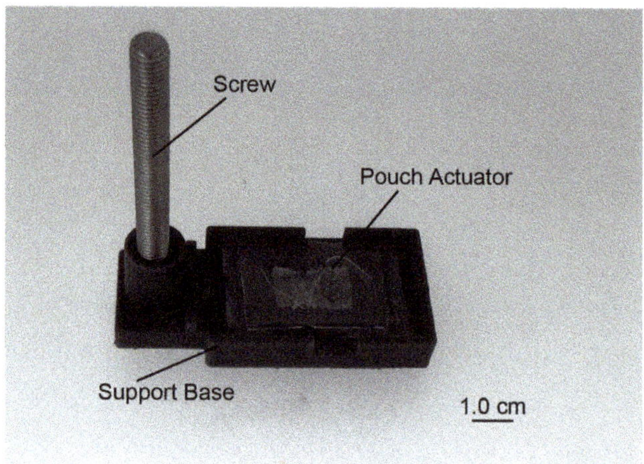

Figure 4. Support base manufactured using 3D printing.

2.3. Detailed Design of Haptic Glove: Areas That Apply Pressure

To replicate a handshake, a suitable pressure needs to be applied to appropriate areas with the right timing. These design parameters were deduced from experiments. First, we examined the contact points and strength distribution in a handshake and selected the position and shape of the pouch on the palm and the back of the hand, as described in Section 2.2. In this experiment, we visualized the contact points between the palm and the back of the hand during a handshake to determine the positions of the pouch actuators. We applied green ink to the palm and fingers of the participants' right hand (Figure 5a). Each participant was asked to shake hands with the other participants. As shown in Figure 5b, after the ink dried, we scanned the palm with a scanner (MFC-L9570CDW, Brother Industries, Nagoya, Japan). We drew the silhouette of the hand on a transparent plastic sheet and placed the hand on it for consistency in the analyses. Five participants, including three males and two females aged 20–25 years, performed two trials each with different partners, which resulted in a total of twenty trials. The images were processed to quantify the contact area.

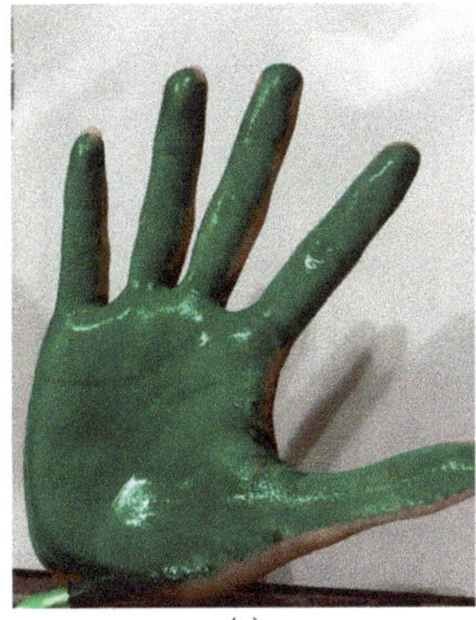
(**a**)

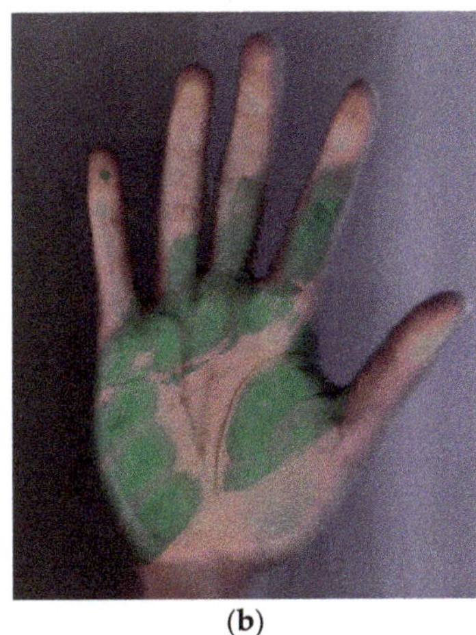
(**b**)

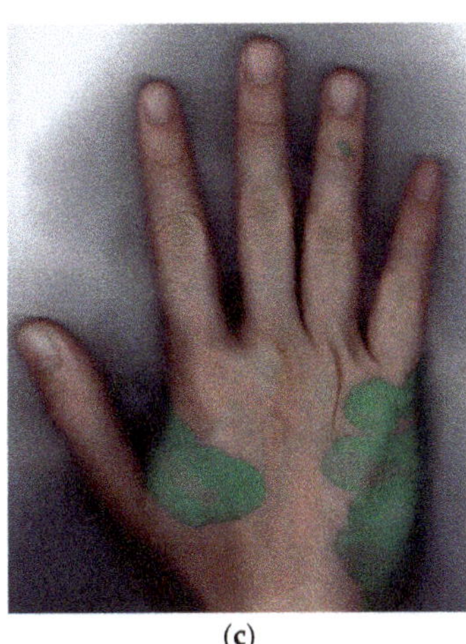
(**c**)

Figure 5. Contact areas during the handshake, as deduced experimentally: (**a**) green ink that was applied to the palm and fingers of each participant's right hand; (**b**,**c**) green ink was transferred to another participant's hand during a handshake.

In the analysis, the hands were normalized as follows (Figure 6). First, point P1 was defined as the interdigital space between the little finger and ring finger, point P2 as the interdigital space between the ring finger and middle finger, and point P3 as the interdigital space between the middle finger and index finger. A straight-line connecting point P1 and point P3 was defined as line m. The hand angle was adjusted so that line m was horizontal. A line perpendicular to line m was drawn from point P2 and this line was defined as line n. The intersection of line n and the contour of the lower bottom of the hand was defined as point P4. To match the aspect ratio to the reference hand, we affine-transformed the entire hand so that line m was 500 mm and line n was 1150 mm in length.

Next, we converted the image from 24-bit RGB to 8-bit grayscale using the NTSC-weighted average method and applied the green filter in Adobe Photoshop CC 2020. Black-and-white binarization was performed using a gray value of 120 as the threshold. A mesh (approximately 6.25 mm $\times$ 6.25 mm, 300 dpi) was set up for each hand, and if more than 50% of each mesh was black, the area corresponding to the mesh was considered to be in contact during the handshake and was labeled as 1. The results of all 20 trials were averaged. This trial was performed for the back of the hand as well.

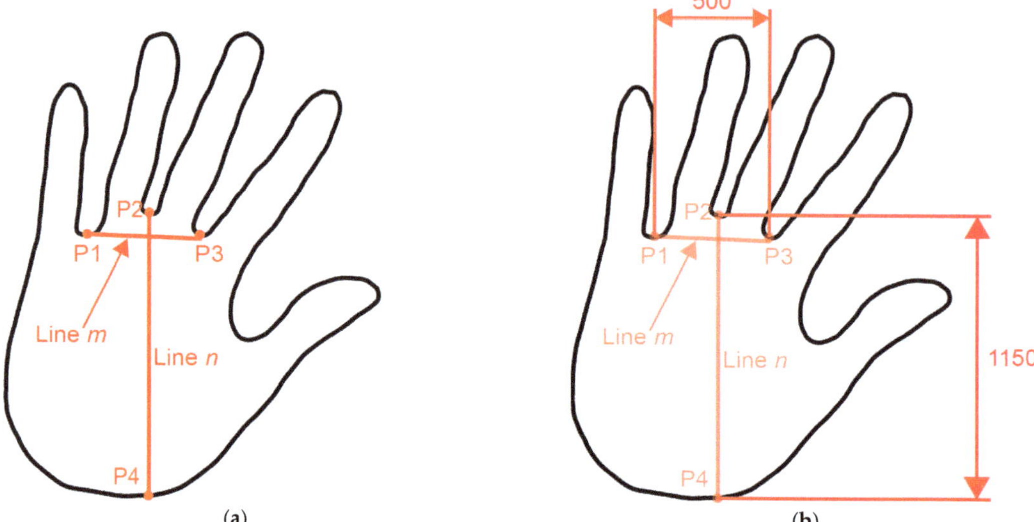

Figure 6. Normalization of palm: (**a**) Points P1–P4 and lines *m* and *n* defined on palm of hand. The angle was adjusted so that line *m* was horizontal. (**b**) Normalization was conducted such that the lengths of lines *m* and *n* were 500 mm and 1500 mm, respectively.

Figure 7 shows the regions where the computed average value was greater than 0.5 among the 20 trials. On the palm of the hand, the three areas of contact during the handshake were referred to as positions I, II and III. On the back of the hand, we found two contact areas, positions IV and V. The liquid-pouch motors, which are positioned in the pocket of the support base, need to cover these areas. The pockets were designed to have a size of 5.0 × 2.0 cm for position I, 4.0 × 2.0 cm for position II and 4.0 × 2.0 cm for position III. On the back of the hand, position IV was 6.0 cm × 2.0 cm and position V was 4.0 cm × 2.0 cm.

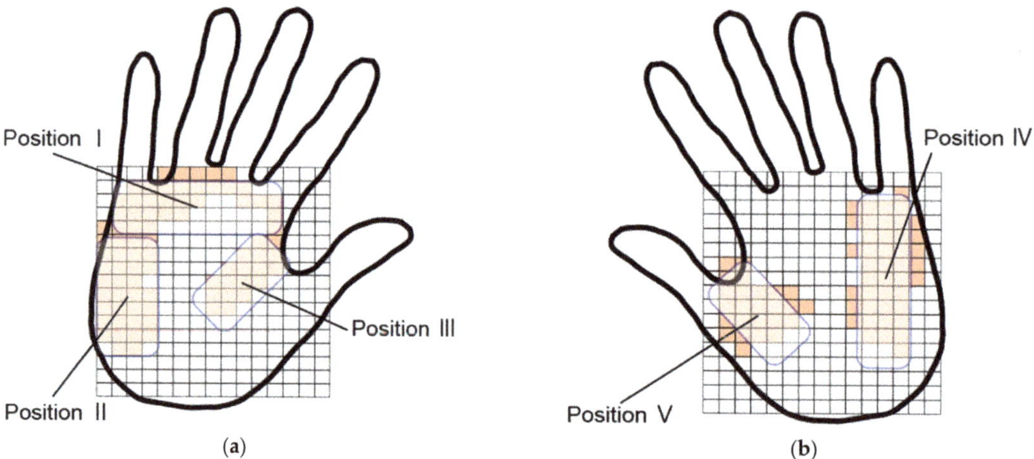

Figure 7. Deduced contact areas during handshake: (**a**) Palm of the hand. (**b**) Back of the hand.

2.4. Detailed Design of Haptic Glove: Distribution of Contact Strength

Next, we experimentally obtained the distribution of the pressure applied to the palm during a handshake, which was needed to determine the requirements of the pouch actuator.

A pressure-sensitive sheet (Prescale, Fujifilm, Tokyo, Japan) was used to obtain the pressure distribution during handshakes. The pressure-sensitive sheet consists of two sheets with a chromophore layer and a developer layer. When sufficient pressure is applied, the microcapsules containing dye in the chromophore layer disintegrate and the dye that leaks from the microcapsules is absorbed by the developer layer and turns red. The range of measurable pressure depends on the size and strength of the microcapsules. The intensity of the color in a particular region represents the pressure.

The following experiment was performed twice on 5 participants (3 males and 2 females between the ages of 20 and 25) for a total of 20 trials. The pressure-sensitive sheets were set in positions I, II and III of each participant's hand and he/she shook hands with another participant. After the handshake, the sheets were collected and scanned (MFC-L9570CDW scanner, Brother Industries, Nagoya, Japan). The color distribution of the sheets was converted to a grayscale image using the green filter correction in Adobe Photoshop CC 2020. Positions I, II and III were divided into grids of 3×8, 3×6 and 3×6 squares, respectively (each measuring approximately 6.25 mm $\times$ 6.25 mm, 300 dpi). As shown in Figure 8, the average gray value for each grid cell was obtained.

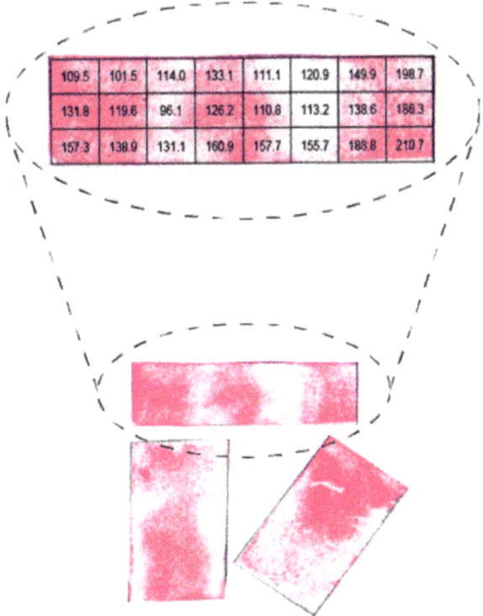

Figure 8. Pressure-sensitive sheets used to deduce the pressure distribution during a handshake. The quantified color intensity represents the pressure distribution in the grid at position I.

Based on these results, the size and position of the pouch actuator were determined to match the pressure distribution for the actuator as close as possible to that for the handshake. The pouch actuators were set on the support base at positions I, II and III. The pressure-sensitive sheets were attached on top of the pouch actuator or the pocket of the haptic glove. The actuators were then heated with an air-dryer to expand and then apply pressure onto the sheets. The pressure-sensitive sheets were scanned as before. Image processing and evaluation were the same as those described above. This method was performed five times for each actuator, and the average gray value for each grid cell was obtained for each actuator.

The pressure applied by the participants and that by the pouch actuators were compared with respect to the magnitude and distribution. The difference was evaluated based on the least-squares error with respect to the position. Figure 9 shows the least-squares error for each actuator at positions I, II and III. Based on the results, the pouch actuators were selected for each position as shown in Table 1. In positions IV and V on the back of the hand, the actuator was often in contact with the tips of the fingers of the other person's hand and the amount of force applied to that position could not be uniquely determined due to individual differences in grip force. Therefore, we decided to install an actuator equivalent to position I for position IV and position III for position V.

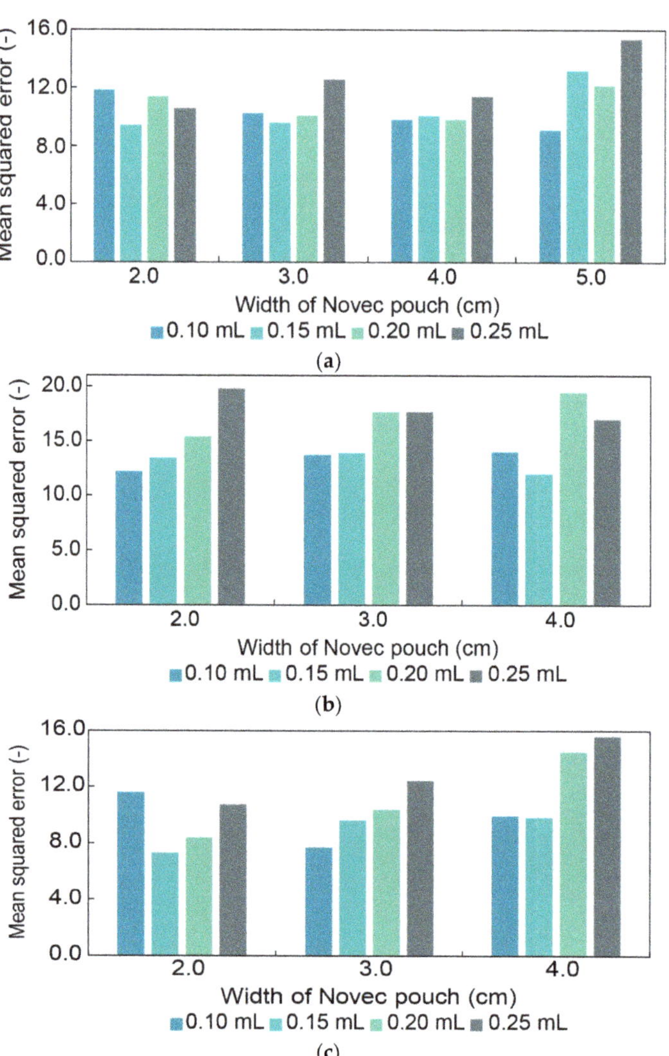

Figure 9. Differences in the pressure distribution between handshakes and pouch actuators at (**a**) position I, (**b**) position II and (**c**) position III.

Table 1. Pouches selected for each position.

Position	Novec ™ 7000 [mL]	Width [cm]
I	0.10	5.0
II	0.15	4.0
III	0.10	2.0

2.5. Detailed Design of the Haptic Glove: Number of Pouch Actuators

The experimental results suggested that three positions (I, II and III) need to be pressed by the appropriately designed pouch actuators to replicate a handshake. However, we surmised that applying pressure to either one or two of these three areas may be sufficient to replicate a handshake. For this preliminary study, a support base measuring 7 cm × 7 cm and covering the entire palm was fabricated. Using this support base, the arrangement of the pouch actuators was experimentally evaluated.

The experiments with one pouch, two pouches and three pouches are referred to here as experiments A, B and C, respectively. The arrangement of pouches for each experiment is shown in Figure 10. In experiment A, a 7 cm × 7 cm pouch with 0.30 mL of Novec ™ 7000 was used (see Figure 10a). Two pouches measuring 5 cm × 2 cm with 0.30 mL of Novec ™ 7000 and measuring 4 cm × 2 cm with 0.30 mL of Novec ™ 7000 were arranged as shown in Figure 10b for experiment B. A maximum of 0.30 mL was encapsulated inside each pouch. We chose the arrangement to cause the largest areas to be covered by the two pouches. For experiment C, three pouches were set to cover positions I, II and III.

(a)

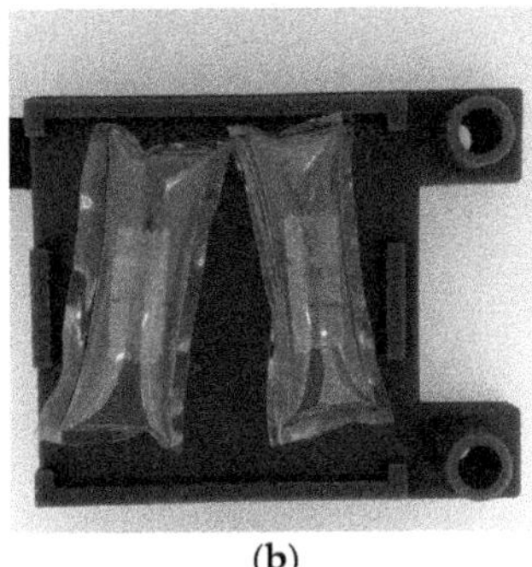
(b)

(c)

Figure 10. Three types of tested haptic gloves: (**a**) 1 liquid-pouch motor, (**b**) 2 motors and (**c**) 3 motors.

First, the pouch actuators were set onto the support base with a pressure-sensitive sheet measuring 7 cm × 7 cm attached to the top. The haptic glove was then set on the hand of a participant and was subsequently heated with a hair dryer. We scanned the pressure-sensitive sheet, normalized the image, binarized it to black and white and rendered it as described in Section 2.4. We conducted this procedure five times for each experiment and deduced the average values for each grid cell. Grid cells with values above 0.5 were considered as the contact points. The results were compared to the results of the handshakes obtained as described in Section 2.4. When a grid cell had the same results, its value was given as 1, otherwise as 0. The sum of these numbers divided by the total number of grid cells was used as the coverage rate to assess the replicability.

Figure 11 shows the results. It was found that experiment C with three actuators exhibited the highest coverage with the lowest standard deviation, though no significant differences were found. We decided to use the haptic glove with three actuators for subsequent experiments to investigate the number of actuators driven out of the three results in the perception of a handshake. Figure 2 shows a photograph of a haptic glove with three liquid-pouch motors.

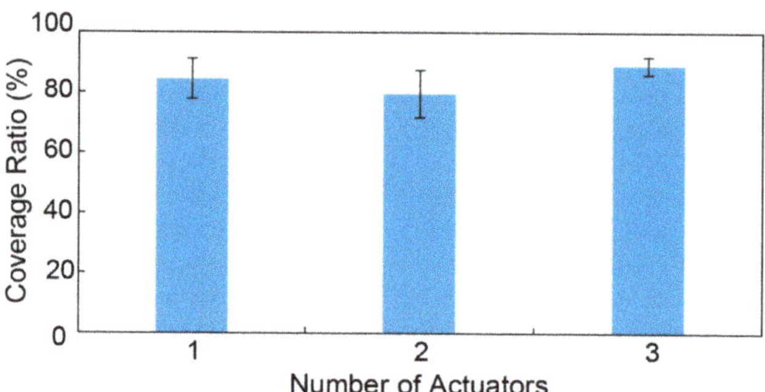

Figure 11. Comparison of coverage for each experiment.

3. Experiments with Proposed Haptic Glove

3.1. Experimental Protocol

The developed haptic glove (see Figure 2) can vary the positions of the three sets of pouch actuator supports independently. This design allowed the device to apply pressure at contact points almost identical to those in a handshake. Seven types of devices were prepared, as shown in Table 2 and Figure 12. The areas where no actuators were present had only a support base, and the pocket was empty. The positions of the actuators on the back of the hand were the same for the seven devices.

Table 2. The seven experimental patterns.

Pattern	Actuators	Number of Actuators
A	Position I	1
B	Position II	1
C	Position III	1
D	Positions I and II	2
E	Positions I and III	2
F	Positions II and III	2
G	All positions	3

A total of 8 participants, 5 males and 3 females aged 20–25 years, participated in the experiment. Figure 13 shows the experiment. The experimental procedure was as follows:

(1) The gloves were heated with a hair dryer (1200 W, Panasonic, Osaka, Japan) for 30 s to ensure that the bag actuators would properly inflate. The temperature of the pouch at that time was appropriate at a translation of 33 °C.

(2) The participant sat on a chair and wore the haptic glove on their right hand.

(3) The participant placed their hand with the device at a predetermined position. A fence prevented participants from seeing their hand.

(4) The gloved hand was heated with a hair dryer from a position about 20 cm away. The actuators were heated over the haptic glove; each actuator was heated for 3 s, and they were actuated sequentially in a clockwise direction for a total of 30 s. Under this condition, none of the 10 participants perceived excessive heat from the dryer. The participants were asked to keep their eyes closed.

(5) The participants were asked to move their hands for 5 s to simulate a handshake. Subsequently, the participants answered survey questions for 30 s. The questions are summarized in Table 3. In addition to the question concerning whether they felt the sensation of a handshake, three additional questions were prepared in this study. The first was whether they felt the force from the actuator. The second was whether they felt warmth from the heated actuator. The third was whether they felt the sensation of being grasped. The participants were asked to respond using a 7-point scale (from 1: Could not feel at all to 7: Feeling was strong). Because the surface temperature of the

palm of the hand was about 33 °C, the pouch actuator did not shrink for about 30 s after the heating stopped.

(6) Steps (1)–(5) were repeated for all glove patterns. Before the next pattern, the tactile glove was removed from the hand and the actuator was allowed to cool down for 1 min. The pouch actuators returned to their original state before the next experiment.

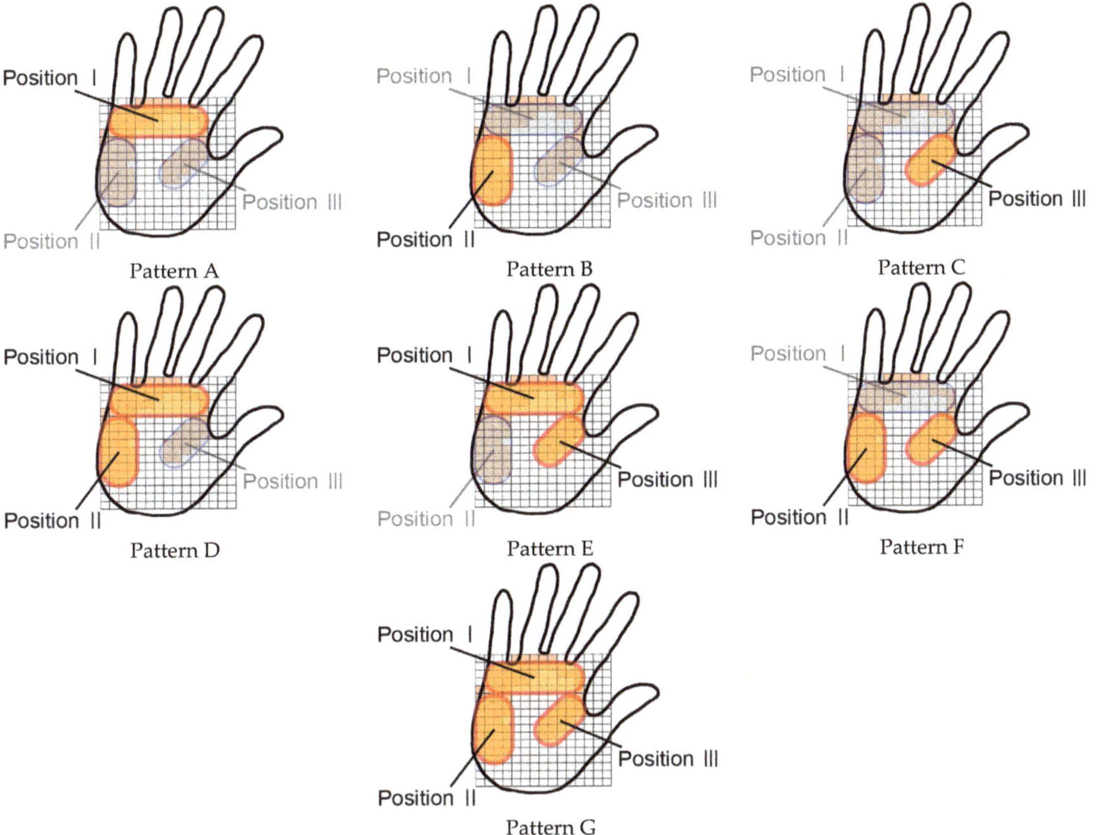

Figure 12. The seven experimental glove patterns.

Table 3. Questions for the sensory evaluation.

No.	Question
1	Did you feel the force?
2	Did you feel warmth like a human body?
3	Did you feel as if you were being grasped?
4	Did you feel as if you were shaking hands?

Figure 13. A photograph of the experiment.

3.2. Experimental Results

Figure 14 reports the means and standard errors of all the evaluated values for each experiment, as well as the significant differences (* $0.01 < p < 0.05$, ** $p < 0.01$).

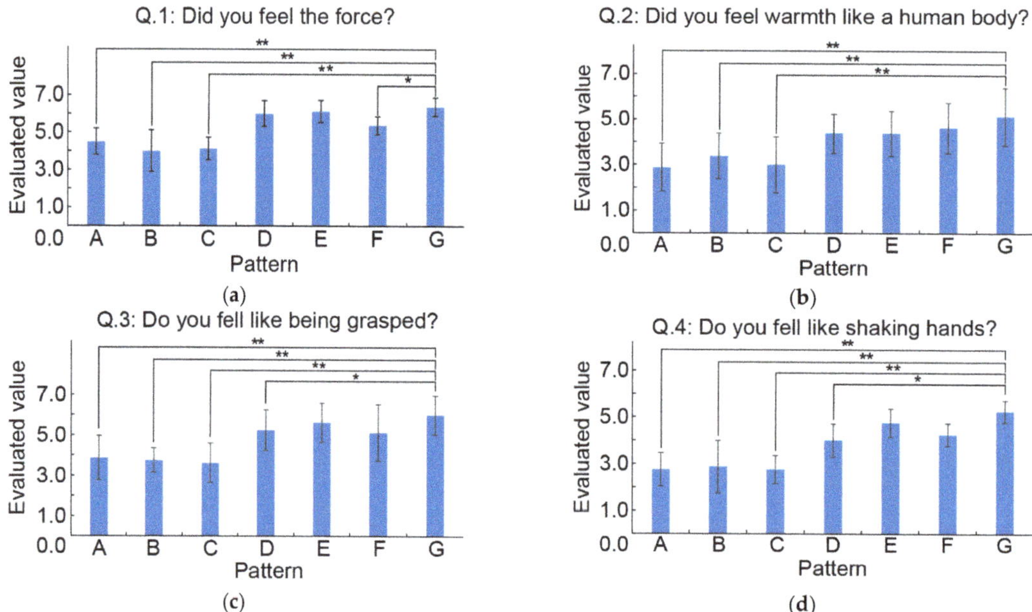

Figure 14. Answers of the participants for questionnaires: (**a**) question 1, (**b**) 2, (**c**) 3 and (**d**) 4. * $0.01 < p < 0.05$, ** $p < 0.01$.

As shown in Figure 14a, glove patterns D, E, F and G with multiple pouch actuators were evaluated higher than patterns A, B and C with a single pouch actuator. Therefore, multiple pouch actuators produced more force than one actuator. Pattern G with three actuators showed a significant difference from A, B, C and F. No significant difference was found among patterns D, E and G. This implies that the actuator in position I plays a crucial role in transferring the force onto the palm. It has been reported that there are many Merkel plates in this position and that they sense delicate skin deformation and are involved in the perception of edges and textures [28]. The other positions, II and III, are known to be less sensitive to force and skin deformation because they have thicker skin and superior deep sensation compared to position I.

Figure 14b shows how the participants perceived the warmth from the actuators. The evaluation value was about 3 for patterns A, B and C with one actuator; 4 for D, E and F with two actuators; and 5 for G with three actuators. We consider that this may originate from the contact areas of the actuators and the palm, which increases with the number of actuators. It is known that there are about 15 cold points and 4 warm points/cm^2 on the palm of the hand, and both cold and warm points are sensitive to temperatures in the range of 15 to 40 °C. The temperature of the actuators is slightly higher than that of the human body and the warm points perceive the warmth.

Figure 14c,d show whether the participants perceived hand holding and handshaking, respectively. The trend was found to be similar: the pattern with a low evaluation for Q3 also had a low evaluation score for Q4. In particular, patterns A, B and C with one actuator had low ratings, and a significant difference from pattern G was obtained. This indicates that sensation from multiple areas is necessary for the participants to perceive the "holding" feeling. The similarity between the results for Q3 and Q4 suggests that the feeling of holding a hand is necessary to feel a handshake. A significant difference was found between patterns D and G in both Q3 and Q4, while patterns E and F did not exhibit a significant difference from G, which implies that the pressure in position III is crucial for perceiving the holding feeling. Position III, the base of the thumb, is in contact with the base of the index finger of the other hand in a handshake. In the experiments with the pressure-sensitive test (see Figure 8), a relatively large force was applied to position III during the handshake.

Based on these results, we consider that the haptic glove with three actuators in positions I, II and III is the best. Handshakes were successfully replicated with a score greater than 5.

3.3. Discussion

Although the proposed haptic glove successfully simulated handshakes with a rating greater than 5, the slow response time of the pouch actuators may make the glove unusable for some applications. For example, in the reproduction of the sense of reality using a combination of vision and touch in a virtual reality environment, tactile reproduction must be timed to the timing of the image and the time delay between tactile and visual stimuli should be less than 1 s [29,30]. The proposed tactile glove uses the heat from a hair dryer to drive the liquid-pouch motors, which takes about 3 s. Some solutions have been already reported, such as direct heating of the pouch with a resistive heater on or in the vicinity of the pouch [24] and infrared-laser irradiation of the pouch actuators with metal particles in the encapsulated liquid [25–27].

In this work, we did not investigate the design of the rigid parts, which are used to efficiently transfer the force from the pouch actuators to the hand. The rigid parts with bone shapes may lead to better replication of a handshake. The rigid parts that are made from soft but harder than the soft actuators can make the whole glove soft and compliant.

4. Conclusions

In this study, we fabricated a prototype haptic glove with gas–liquid phase-change actuators or liquid-pouch motors and successfully demonstrated simulated handshakes. The liquid-pouch motors are light and soft and do not require external devices, such as pneumatic valves or batteries, which improves the convenience of their use. The design requirements for the glove, namely, the position, number and force output of the pouch actuators, were experimentally determined. Perception experiments validated the effectiveness of the proposed haptic glove in simulating a handshake, which was scored by the participants as more than five out of seven points. This prototype has revealed that it provides the feedback of a handshake sensation. It is expected that further effects can be expected in the future by improving the heating method and the material of the support.

Supplementary Materials: The following supporting information can be downloaded at: https://www.mdpi.com/article/10.3390/act12020051/s1, Video S1: Simulated handshake.

Author Contributions: Conceptualization, S.Y. and N.M.; methodology, S.Y. and N.M.; software, S.Y.; validation, S.Y., T.H., H.I. and N.M.; formal analysis, S.Y.; investigation, T.H., H.I. and N.M.; resources, N.M.; data curation, S.Y. and N.M.; writing—original draft preparation, S.Y.; writing—review and editing, N.M.; visualization, S.Y. and N.M.; supervision, T.H., H.I. and N.M.; project administration, N.M.; funding acquisition, N.M. All authors have read and agreed to the published version of the manuscript.

Funding: This work was supported by JSPS KAKENHI Grant Number 20H02121 and JST CREST Grant Number JPMJCR19A2, Japan.

Data Availability Statement: The data presented in this study are openly available in FigShare at https://doi.org/10.6084/m9.figshare.21769196. Accessed on 22 December 2022.

Conflicts of Interest: The authors declare no conflict of interest.

References

1. Graham, W.; Thomas, C.; Sriram, S.; Stephen, A.B. Perception of ultrasonic haptic feedback on the hand: Localisation and apparent motion. In *CHI '14: Proceedings of the SIGCHI Conference on Human Factors in Computing Systems, Toronto, ON, Canada, 26 April–1 May 2014*; Association for Computing Machinery: New York, NY, USA, 2014; pp. 1133–1142.
2. Wilder, P.; Edwin, B. Somatic motor and sensory representation in the cerebral cortex of man as studied by electrical stimulation. *Brain* **1937**, *60*, 389–443.
3. Greg, L.S.; Susan, L.D.; Murray, R.B.; Todd, C.D. Exploring the handshake in employment interviews. *J. Appl. Psychol.* **2008**, *93*, 1139–1146.
4. Chaplin, W.F.; Phillips, J.B.; Brown, J.D.; Clanton, N.R.; Stein, J.L. Handshaking, gender, personality, and first impressions. *J. Pers. Soc. Psychol.* **2000**, *79*, 110–117. [PubMed]
5. Nunez, C.M.; Huerta, B.N.; Okamura, A.M.; Culbertson, H. Investigating Social Haptic Illusions for Tactile Stroking (SHIFTS). In Proceedings of the 2020 IEEE Haptics Symposium (HAPTICS), Crystal City, VA, USA, 28–31 March 2020; Volume 1, pp. 629–636.
6. Takahashi, N.; Okazaki, R.; Okabe, H.; Yoshikawa, H.; Aou, K.; Yamakawa, S.; Yokoyama, M.; Kajimoto, H. Sense-roid: Emotional haptic communication with yourself. In Proceedings of the Virtual Reality International Conference (VRIC 2011), Laval, France, 6–8 April 2011; Volume 1, pp. 6–8.
7. Mourad, B.; Grigore, B.; Grigore, P.; Rares, B. The Rutgers Master II—New Design ForceFeedback Glove. *IEEE/ASME Trans. Mechatron.* **2002**, *7*, 256–263.
8. Kitazaki, M.; Hamada, T.; Yoshiho, K.; Kondo, R.; Amemiya, T.; Hirota, K.; Ikei, Y. Virtual walking sensation by prerecorded oscillating optic flow and synchronous foot vibration. *IEEE Comput. Graph. Appl.* **2018**, *38*, 44–56. [CrossRef]
9. Hirose, M.; Hirota, K.; Ogi, T.; Yano, H.; Kakehi, N.; Saito, M.; Nakashige, M. HapticGEAR: The development of a wearable force display system for immersive projection displays. In Proceedings of the IEEE Virtual Reality 2001, Yokohama, Japan, 13–17 March 2001; pp. 123–129.
10. Talhan, A.; Kim, H.; Jeon, S. Tactile Ring: Multi-Mode Finger-Worn Soft Actuator for Rich Haptic Feedback. *IEEE Access* **2020**, *8*, 957–966. [CrossRef]
11. Wang, Z.; Wang, D.; Zhang, Y.; Liu, J.; Wen, L.; Xu, W.; Zhang, Y. A three-fingered force feedback glove using fiber-reinforced soft bending actuators. *IEEE Trans. Ind. Electron.* **2020**, *67*, 7681–7690.
12. Uddin, M.W.; Zhang, X.; Wang, D. A pneumatic-driven haptic glove with force and tactile feedback. *Front. Bioeng. Biotechnol.* **2017**, 304–311. [CrossRef]
13. Premarathna, C.P.; Chathuranga, D.S.; Lalitharatne, T.D. Fabrication of a soft tactile display based on pneumatic balloon actuators and voice coils: Evaluation of force and vibration sensations. In Proceedings of the SII 2017—2017 IEEE/SICE International Symposium on System Integration, Taipei, Taiwan, 11–14 December 2017; pp. 763–768.
14. Rognon, C.; Koehler, M.; Duriez, C.; Floreano, D.; Okamura, A.M. Soft Haptic Device to Render the Sensation of Flying Like a Drone. *IEEE Robot. Autom. Lett.* **2019**, *4*, 2524–2531. [CrossRef]
15. Delazio, A.; Nakagaki, K.; Lehman, J.F.; Klatzky, R.L.; Sample, A.P.; Hudson, S.E. Force jacket: Pneumatically-actuated jacket for embodied haptic experiences. In Proceedings of the Conference on Human Factors in Computing Systems, Montreal, QC, Canada, 21–26 April 2018.
16. Bardaweel, H.K.; Anderson, M.J.; Weiss, L.W.; Richards, R.F.; Richards, C.D. Characterization and modeling of the dynamic behavior of a liquid-vapor phase change actuator. *Sens. Actuators A Phys.* **2009**, *149*, 284–291. [CrossRef]

17. Henning, A.K. Liquid and gas-liquid phase behavior in thermopneumatically actuated microvalves. *Microfluid. Devices Syst.* **1998**, *3515*, 53.
18. Akagi, T.; Dohta, S.; Fujimoto, S.; Tsuji, Y.; Fujiwara, Y. Development of flexible thin actuator driven by low boiling point liquid. *Dep. Intell. Mech. Eng.* **2015**, *3*, 55–59. [CrossRef]
19. Tsuji, Y.; Dohta, S.; Akagi, T.; Fujiwara, Y. Analysis of flexible thin actuator using gas-liquid phase-change of low boiling point liquid. In Proceedings of the 3rd International Conference on Intelligent Technologies and Engineering Systems (ICITES2014), Kaohsiung, Taiwan, 19–21 December 2014; 2016; Volume 345, pp. 67–73.
20. Nakahara, K.; Narumi, K.; Niiyama, R.; Kawahara, Y. Electric phasechange actuator with inkjet printed flexible circuit for printable and integrated robot prototyping. In Proceedings of the IEEE International Conference on Robotics and Automation, Singapore, 29 May–3 June 2017; pp. 1856–1863.
21. Narumi, K.; Sato, H.; Nakahara, K.; Seong, A.Y.; Morinaga, K.; Kakehi, Y.; Niiyama, R.; Kawahara, Y. Liquid pouch motors: Printable planar actuators driven by liquid-to-gas phase change for shape-changing interfaces. *IEEE Robot. Autom. Lett.* **2020**, *5*, 3915–3922.
22. Sanchez, V.; Payne, J.C.; Preston, J.D.; Alvarez, T.J.; Weaver, C.J.; Atalay, T.A.; Boyvat, M.; Daniel, M.; Vogt, M.D.; Wood, J.R.; et al. Smart thermally actuating textiles. *Adv. Mater. Technol.* **2020**, *5*, 200038. [CrossRef]
23. Hirai, S.; Nagatomo, T.; Hiraki, T.; Ishizuka, H.; Kawahara, Y.; Miki, N. Micro Elastic Pouch Motors: Elastically Deformable and Miniaturized Soft Actuators Using Liquid-to-Gas Phase Change. *IEEE Robot. Autom. Lett.* **2021**, *6*, 5373–5380. [CrossRef]
24. Bell, L.E. Cooling, heating, generating power, and recovering waste heat with thermoelectric systems. *Science* **2008**, *321*, 1457–1461. [CrossRef]
25. Altmüller, R.; Schwödiauer, R.; Kaltseis, R.; Bauer, S.; Graz, I.M. Large area expansion of a soft dielectric membrane triggered by a liquid gaseous phase change. *Appl. Phys. A Mater. Sci. Process.* **2011**, *105*, 1–3.
26. Usui, T.; Ishizuka, H.; Hiraki, T.; Kawahara, Y.; Ikeda, S.; Oshiro, O. Miniaturized Liquid Pouch Motors Using Flexible Liquid Metal Heater. In Proceedings of the 33rd International Microprocesses and Nanotechnology Conference (MNC 2020), Online, 9–12 November 2020; Volume 3922, pp. 1–2.
27. Meder, F.; Naselli, G.A.; Sadeghi, A.; Mazzolai, B. Remotely Light-Powered Soft Fluidic Actuators Based on Plasmonic-Driven Phase Transitions in Elastic Constraint. *Adv. Mater.* **2019**, *31*, 1905671.
28. Kobayashi, K.; Maeno, T. Relationship between the Structure of Finger Tissue and the Location of Tactile Receptors. *JSME Int. J. Ser. C Mech. Syst. Mach. Elem. Manuf.* **1998**, *41*, 94–100.
29. Kameyama, S.; Ishibashi, Y. Perception of synchronization errors in haptic and visual communications. In Proceedings of the Multimedia Systems and Applications IX, Boston, MA, USA, 2–3 October 2006; pp. 67–74.
30. Silva, M.J.; Orozco, M.; Cha, J.; Saddik, E.A.; Petriu, M.E. Human perception of haptic-to-video and haptic-to-audio skew in multimedia applications. *ACM Trans. Multimed. Comput. Commun. Appl.* **2013**, *9*, 1–16. [CrossRef]

Article

An Origami-Inspired Negative Pressure Folding Actuator Coupling Hardness with Softness

Zhaowen Shao [1], Wentao Zhao [1], Zhaotian Zuo [1], Jun Li [1,*] and I-Ming Chen [2]

[1] Ministry of Education Key Laboratory of Measurement and Control of CSE, Southeast University, Nanjing 210096, China
[2] School of Mechanical and Aerospace Engineering, Nanyang Technological University, Singapore 639798, Singapore
* Correspondence: j.li@seu.edu.cn

Abstract: Soft actuators have a high potential for the creative design of flexible robots and safe human–robot interaction. So far, significant progress has been made in soft actuators' flexibility, deformation amplitude, and variable stiffness. However, there are still deficiencies in output force and force retention. This paper presents a new negative pressure-driven folding flexible actuator inspired by origami. First, we establish a theoretical model to predict such an actuator's output force and displacement under given pressures. Next, five actuators are fabricated using three different materials and evaluated on a test platform. The test results reveal that one actuator generates a maximum pull force of 1125.9 N and the maximum push force of 818.2 N, and another outputs a full force reaching 600 times its weight. Finally, demonstrative experiments are conducted extensively, including stretching, contracting, clamping, single-arm power assistance, and underwater movement. They show our actuators' performance and feature coupling hardness with softness, e.g., large force output, strong force retention, two-way working, and even muscle-like explosive strength gaining. The existing soft actuators desire these valuable properties.

Keywords: flexible actuator; origami; folding structure; output force; force retention; two-way locomotion; explosive force

Citation: Shao, Z.; Zhao, W.; Zuo, Z.; Li, J.; Chen, I.-M. An Origami-Inspired Negative Pressure Folding Actuator Coupling Hardness with Softness. *Actuators* **2023**, *12*, 35. https://doi.org/10.3390/act12010035

Academic Editor: Hamed Rahimi Nohooji

Received: 30 November 2022
Revised: 2 January 2023
Accepted: 4 January 2023
Published: 10 January 2023

1. Introduction

Rigid robots are difficult to meet the increasing requirements for flexible operation due to their inherent deficiencies in flexibility and adaptability. In the past decade, people continuously studied a variety of soft robots composed of flexible materials. They have natural flexibility and adaptability for flexible operations. Therefore, soft actuators have attracted more attention as power components for developing soft robots.

A wide variety of soft actuation systems have been reported in the literature, e.g., cable-driven, fluid-driven, shape memory materials (SMMs), electroactive polymers (EAPs), and magneto- and electro-rheological materials (M/E-RMs) [1–3]. A cable-driven actuation system powers a soft robot by pulling a cable fixed on the robot's main body with the help of a motor [4–7]. This method can offer low inertia, fast response, and high output force according to the selected motor. SMMs can plastically deform into temporary shapes and recover the original shapes using thermal stimulation. The cycle can always be repeated. SMMs can be divided into two categories: shape memory alloys (SMAs) [8–11] and shape memory polymers (SMPs) [12]. The materials switch back and forth between two temperatures, so the operating frequency of such an actuation manner is low. When an electric field is applied, the EAP shows a change in size or shape, resulting in actuation. Two EAPs are used in the field of soft robots, i.e., dielectric elastomer actuators (DEAs) [13–16] and ionic polymer metal composites (IPMCs) [17–22]. DEA has a high response frequency but requires a high voltage, generally greater than 1kV. IPMC works under low voltage, but the output force is small. A M/E-RM actuation system embedded with electric or magnetic particles requires an external electric or magnetic field to work. M/E-RM, including magnetorheological materials (MRM) [23–26] and electrorheological materials (ERM) [27], is suitable for making soft microactuators.

Compared with the above soft actuators, fluid-driven ones have prominent superiorities in deformation amplitude and output force [28,29]. Therefore, they have been the research focus in recent years. Some researchers use hyperelastic materials, such as silica gel or foam, to make soft actuators [30–37]. These actuators work via the cumulative deformation caused by the expansion of silica gel material under positive pressure. The maximum output force they can produce is about 1.2~15.2N and primarily be used to produce large deformation.

On the other hand, some studies reinforce soft actuators by adding fibers to silica gel materials [38–40]. The maximum output force of these actuators can be up to 25~40 N. However, it limits, to a great extent, the actuator's deformation range. In addition, some soft actuators are made of fiber materials [41–47]. Their maximum output force is about 45~320 N. Compared with actuators based on hyperelastic materials, such actuators can bear more significant pressure and produce greater output force, but their deformation amplitude is much smaller. Further, to achieve a greater output force, some work uses rigid folding skeletons sealed with a tough impermeable film to form soft actuators [48–52]. As a result, the output force generated can reach 90~630 N, but the skeletons severely limit their deformation range.

The fluid-driven soft actuators mentioned above, regardless of material and structure, almost have a common feature, i.e., the output force decreases significantly with the increase of displacement. Therefore, it leads to poor force retention during work. Specifically, their output force is maximal at the initial position and decays rapidly with increasing displacement under constant pressure. Therefore, it may result in accidental situations, such as state instability and damage.

Some studies improve the retention ability of soft actuators with variable stiffness during work [53–59]. However, variable stiffness can only maintain some specific positions statically. At the same time, variable stiffness may restrict the actuator's motion capability and output force.

In summary, substantial progress has been made in promoting the output force of soft actuators, but their force retention is still far from enough. However, soft actuators must have either a large output force or strong force retention for some flexible operations.

In this paper, we propose a new soft actuator paradigm with a folding skeleton and a sealed skin. Sealed by a soft film, it can be driven by negative pressure air or liquid and work in two ways. Furthermore, its output force increases significantly with the increase of displacement and can reach infinity in theory.

The main contributions of this work are as follows.

(1) Inspired by origami, we propose a novel two-way stretching flexible actuator structure coupling hardness with softness. The negative-pressure fluid-driven actuator has strong force retention and large output force simultaneously. Additionally, it can produce explosive force as human muscles do. We fabricate five actuators with a patched, 3D-printed, or machined skeleton. The test results reveal that one generates a maximum pull force of 1125.9 N and the maximum push force of 818.2 N, and another's full output force reaches 600 times its weight.

(2) We conduct several innovative demonstrations of the actuators, including stretching, contracting, clamping, single-arm power assistance, and underwater movement. They show our actuators' characteristics of a large output force, strong force retention, two-way working, and even generating explosive force as human muscles do. The existing soft actuators desire these valuable properties. In addition, the demonstrations indicate the excellent application potential of our flexible actuators.

The rest of this paper is organized as follows. In Section 2, materials and methods are introduced, including the working principle, structural design and mechanical analysis, and fabrication of our actuators. In Section 3, extensive experiments are conducted, and the results are analyzed and discussed. Finally, the main conclusions are drawn, and future work is planned in Section 4.

2. Materials and Methods

2.1. Working Principle

We propose a folding structure, as shown in Figure 1a. It comprises many vertical and horizontal plates connected by a hard or soft hinge. The rotation of a vertical plate can drive the sliding of the horizontal plate. We lay out multiple vertical plates symmetrically and enhance the movement effect through the synergy of the plates.

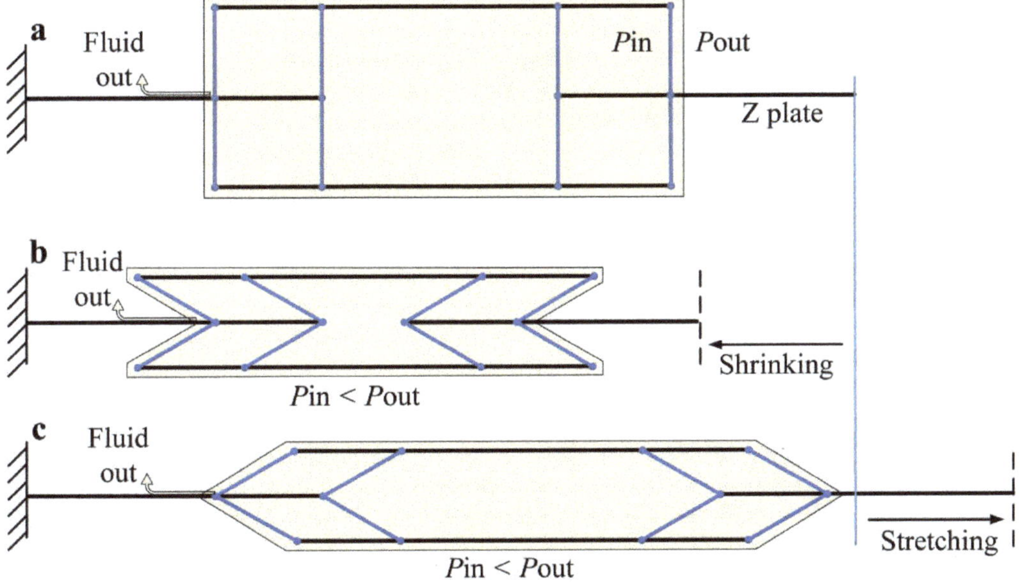

Figure 1. The folding structure's state evolvement: (**a**) A flexible actuator in the initial state; (**b**) The flexible actuator is shrinking; (**c**) The flexible actuator in stretching.

To form an airtight flexible actuator, we seal this structure with a soft film and drive it through negative pressure. The state evolvement of this actuator under negative pressure fluid is illustrated in Figure 1. Under negative pressure, the actuator can work in two ways, i.e., shrinking and stretching. They depend on the actuator's initial state, which is decided by the angle between the vertical and horizontal plates. The actuator's stretching generates a push force, and shrinking produces a pull force, as shown in Figure 1b,c.

Note that because the actuator's deformation mode depends on its initial state determined by external forces, it cannot work simultaneously as an extendable and retractable actuator.

2.2. Structural Design and Mechanical Analysis

The initial 3D folding structure is rendered as illustrated in Figure 2a. When an actuator of this structure works under negative pressure, the soft film greatly concaves to the inside. As a result, it significantly reduces the output force of the actuator. To minimize such a negative impact, we add side supports to the structure, as shown in Figure 2b.

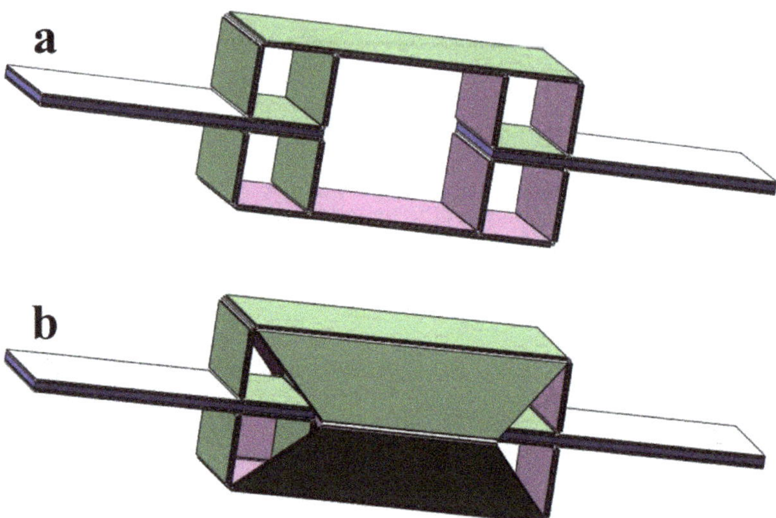

Figure 2. Origami-inspired 3D folding structures: (**a**) Initial configuration; (**b**) Improved configuration with side support.

The forces on an actuator with the improved skeleton driven by negative pressure are shown in Figure 3a–d, including F_{N2} on the side support, F_{N1}, F_{N2r}, F_{N3}, F_{N4}, and F_A on the mainframe when shrinking or stretching. L_1 and L_4 represent the length of each vertical and horizontal plate, respectively. L_2 represents the distance between the rotating shafts at both ends of the beam, and L_b represents the length of the beam itself. Symbol h represents the width of each vertical plate and horizontal plate of the skeleton, and L_3 is the width of the side supporting structure. Additionally, d_0 denotes the center distance of two adjacent hinges on both sides of plate Z, and L_0 is the distance between the outer edge of the upper plate and the axis of the adjacent hinge.

We assume that the flexible film is well-sealed, and its deformation has a negligible effect on the skeleton. The beam's position in the actuator is shown in Figure 3b,c. In the skeleton, the stiffness of the beam is relatively insufficient, and elastic deformation is more likely to occur. Therefore, we first analyze the deformation of the beam, as shown in Figure 3d.

We assume that (1) we only consider the elastic deformation of the beam and treat other parts in the skeleton as rigid bodies, (2) the beam bends without changing its length, and (3) the beam is in a static equilibrium state after deformation and regarded as a rigid body.

As shown in Figure 3d, the beam deforms and maintains balance under the combined action of load concentration q and the connecting structures at both ends. The load concentration q is determined by the pressure p and the beam's width h, that is,

$$q = ph \tag{1}$$

Since the beam's deformation in length is not considered, the load at both ends is equivalent to a fixed support and movable support, as shown in Figure 3d.

The bending moment equation of the beam's section at position x is,

$$M_z(x) = \frac{1}{2}qL_b x - \frac{1}{2}qx^2 \tag{2}$$

The approximate differential equation of the deflection curve of the beam is,

$$\frac{d^2v}{dx^2} = \frac{M_z(x)}{EI_z} \tag{3}$$

where E is the material's elastic modulus, I_z is the inertia moment of the beam, and EI_z is the bending stiffness of the beam. v is the deflection of the beam at position x, and α is the corner of the section at position x.

In the case of small deformation, the relation between the angle α of the beam and the deflection v is,

$$\alpha \approx tan\alpha = \frac{dv}{dx} \tag{4}$$

According to Equations (2) and (3), we can have the following equation.

$$EI_z\frac{d^2v}{dx^2} = M_z(x) = -\frac{1}{2}qx^2 + \frac{1}{2}qL_bx \tag{5}$$

We integrate the two sides of Equation (5) two times into:

$$EI_z\alpha = \int M_z(x)dx = -\frac{1}{6}qx^3 + \frac{1}{4}qL_bx^2 + C \tag{6}$$

$$EI_zv = \int [\int M_z(x)dx]dx = -\frac{1}{24}qx^4 + \frac{1}{12}qL_bx^3 + Cx + D \tag{7}$$

where C and D are constants.

The boundary conditions of the left endpoint A and the middle point of the beam are, respectively,

$$\alpha|_{x=\frac{L_b}{2}} = 0 \tag{8}$$

$$v|_{x=0} = 0 \tag{9}$$

We substitute Equations (8) and (9) into Equations (6) and (7), respectively, and obtain the following results,

$$C = \frac{1}{24}qL_b^3 \tag{10}$$

$$D = 0 \tag{11}$$

Therefore,

$$\alpha = \frac{1}{EI_z}\left(-\frac{1}{6}qx^3 + \frac{1}{4}qL_bx^2 + \frac{1}{24}qL_b^3\right) \tag{12}$$

$$v = \frac{1}{EI_z}\left(-\frac{1}{24}qx^4 + \frac{1}{12}qL_bx^3 + \frac{1}{24}qL_b^3x\right) \tag{13}$$

According to Equation (12), we can obtain the following maximum bending angle of the beam,

$$\alpha_{max} = \alpha|_{x=0} = \alpha|_{x=L_b} = \frac{1}{24}qL_b^3 \tag{14}$$

From Equation (14), we can find that the beam's bending deformation is mainly related to the pressure p. The greater the pressure p, the greater the bending amplitude.

The beam after deformation is an approximately circular arc. Therefore, the linear distance between A and B at both ends of the beam after deformation is,

$$L_b' = \frac{\sin\alpha_{max}}{\alpha_{max}}L_b \tag{15}$$

There is a similar beam in the side support structure, its deformation is also consistent with those of the upper beam.

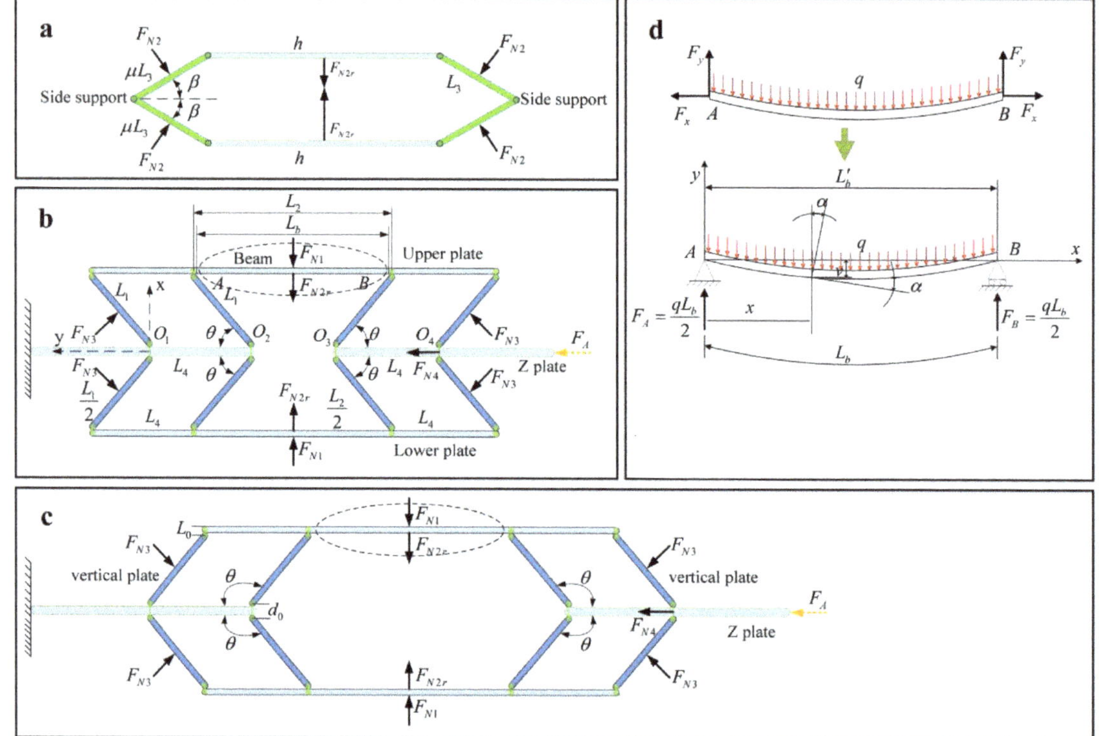

Figure 3. Force analysis of the actuator: (**a**) Force on the side support; (**b**) Force on the mainframe when shrinking; (**c**) Force on the mainframe when stretching; (**d**) Beam deformation analysis.

Then, we calculate the equivalent concentrated forces on every plate on the skeleton. Pressure differences between the actuator's inside and outside cause them. The equivalent concentrated force of the uniformly distributed load on different plates on the skeleton is given by,

$$F_{N1} = pS_{N1} - ph\left(L_b - L_b'\right) \tag{16}$$

$$F_{N2} = pS_{N2} - pL_3\left(L_b - L_b'\right) \tag{17}$$

$$F_{Ni} = pS_{Ni},\ i = 3,\ 4 \tag{18}$$

where S_{N1} is the area of an upper plate or lower plate, S_{N2} is the area of a single plate on the side support, S_{N3} is the area of a vertical plate, and S_{N4} is the equivalent area of the uniformly distributed load on the single Z plate. F_{N1}, F_{N2}, F_{N3}, and F_{N4} are the equivalent concentrated forces of the uniformly distributed loads on the above plates, as shown in Figure 3.

According to Figure 3a, we convert F_{N2} on the side into its equivalent concentrated force F_{N2r} on the mainframe as follows,

$$F_{N2r} = F_{N2}\frac{\mu - \sin^2\beta}{\cos\beta} \tag{19}$$

In Equation (18), μ is the coefficient of the arm of force F_{N2r} in the above conversion, and its value is determined by the size of the side supporting structure, as shown in Table 1. β is half of the angle between two adjacent side supporting plates, as shown in Figure 3a.

θ is the angle between the vertical plate and the Z plate, as shown in Figure 3b,c. The relationship between angles θ and β can be denoted by,

$$\beta = \arcsin\frac{d_0 + 2L_0 + 2L_1 \sin\theta}{2L_3} \tag{20}$$

Table 1. Basic parameters of different actuators.

Actuators	L_1 (mm)	L_2 (mm)	L_b (mm)	L_3 (mm)	L_4 (mm)	d_0 (mm)	L_0 (mm)	h (mm)	μ	S_{N1} (mm)	S_{N2} (mm)	S_{N3} (mm)	S_{N4} (mm)
Patched skeleton	22	52	52	27	22	3	0	42	0.55	4032	2025	924	0
Machined skeleton	22	54	49	40	32	14	6	42	0.54	5875	3844	630	882
3D-printed skeleton	22	54	49	40	32	14	6	42	0.54	5875	3844	630	882

Therefore, the total equivalent force on either the upper or lower plate on the skeleton is,

$$F_N = F_{N1} + F_{N2r} \tag{21}$$

Assume that the direction of the output force is positive when the actuator folds. Then, according to the force analysis shown in Figure 3b,c, force F_t generated by the unilateral skeleton can be calculated by,

$$F_t = \frac{\frac{1}{2}(F_{N1} + F_{N2r})\cos\theta - \frac{1}{2}F_{N3}}{\sin\theta} + F_{N3}\sin\theta \tag{22}$$

Thus, we have the total force F_A generated by the pressure difference between the inner and outer sides of the skeleton,

$$F_A = 2F_t + F_{N4} \tag{23}$$

Further, by substituting Equations (1) and (14)–(22) into Equation (23), we can obtain the following result,

$$\begin{aligned} F_A &= \frac{(F_{N1}+F_{N2r})\cos\theta - F_{N3}}{\sin\theta} + 2F_{N3}\sin\theta + F_{N4} \\ &= p\left[\frac{\left(S_{N1} - h\left(L_b - L_b'\right) + \left[S_{N2} - L_3\left(L_b - L_b'\right)\right]\frac{\mu - \sin^2\beta}{\cos\beta}\right)\cos\theta - S_{N3}}{\sin\theta} + 2S_{N3}\sin\theta + S_{N4}\right] \end{aligned} \tag{24}$$

By Equation (24), the output force of the actuator has a nearly linear relationship with the pressure difference p between its inner and outer sides. In contrast, it has a nonlinear relationship with angle θ. As the actuator shrinks, the angle θ gradually decreases while the output force gradually increases. When the actuator shrinks to the limit position, θ approaches zero and the output force of the actuator tends to infinity. Similarly, when the actuator is in the extended state, the angle θ gradually increases while the output force gradually increases, and its maximum output force tends to infinity.

We consider the initial state of the actuator is when the vertical plate is perpendicular to the horizontal plate, or $\theta = 90°$. Assume that the direction of the displacement S is positive when the actuator folds, as shown in Figure 3b.

The initial distance between endpoints O_1 and O_4 on Z plates at the left and right sides of the skeleton is,

$$D_o = L_2 + 2L_4 \tag{25}$$

During the action of the actuator, the beam bends, and the linear distance L_b between its two endpoints, A and B, becomes L_b'. Correspondingly, the linear distance L_2 between the rotating shafts at both ends of the beam turns into L_2'. Then, we have,

$$L_2 - L_2' = L_b - L_b' \tag{26}$$

Therefore, the distance between endpoints O_1 and O_4 is,

$$D = L_2' + 2L_4 - 2L_1 \cos\theta \tag{27}$$

The displacement S of the actuator can be expressed as,

$$S = D_o - D = \frac{\alpha_{\max} - \sin\alpha_{max}}{\alpha_{max}} L_b + 2L_1 \cos\theta \tag{28}$$

By substituting Equations (1) and (14) into Equation (28), we obtain the displacement S.

$$S = \frac{\frac{1}{24}phL_b^3 - \sin\left(\frac{1}{24}phL_b^3\right)}{\frac{1}{24}phL_b^3} L_b + 2L_1 \cos\theta \tag{29}$$

By Equation (29), the actuator's displacement is mainly related to θ. Specifically, $0^\circ < \theta < 90^\circ$ if the actuator contracts and S increases as θ decreases. If the actuator elongates, $90^\circ < \theta < 180^\circ$ and S increases as θ increases.

The limit displacement of the actuator is two times of L_1, no matter whether it is retracted or extended, namely,

$$S_{max} = 2L_1 \tag{30}$$

2.3. *Fabrication of Actuator Prototypes*

An actuator's performance and features are determined mainly by its material and fabrication process. Different actuators can be used for different purposes. We fabricate five actuators using several common manufacturing processes and materials. Each has a rigid skeleton supporting the actuator and conducting force output. The patched, machined, or 3D-printed skeleton is sealed by an airtight soft film with good toughness to form a sealed cavity.

2.3.1. Patched Actuator

The fabrication process of the actuator with a patched skeleton is shown in Figure 4. We use a 0.5 mm thick white canvas as the soft substrate.

First, we paste several 0.5 mm thick polyethylene terephthalate (PET) patches and 0.6 mm thick carbon fiber plate (CFP) patches onto the canvas' two sides' specific positions using glue. Then, we get a combination through cutting, folding, pasting, etc., as shown in Figure 4a.

Next, we obtain the skeleton through subsequent operations, as shown in Figure 4b.

Finally, the skeleton is sealed with a 0.1 mm thick thermoplastic polyurethane (TPU) film to form an actuator, shown in Figure 4c.

The actuator we made with the above method is shown in Figure 5a. The weight of this actuator is 76.5 g, and the overall size is 220 × 68 × 52 mm, and its basic dimension parameters are in Table 1.

2.3.2. Machined Actuator

Considering that the patched actuator's fabrication process is too cumbersome and overall rigidity is insufficient, we use aluminum alloy to make the skeleton. Specifically, we machine the individual aluminum alloy plates and assemble them into a skeleton. Then, we seal the skeleton with a 0.1 mm thick TPU film and obtain an actuator, shown in Figure 5b. Its weight is 462 g, its total size is 240 × 78 × 78 mm, and its dimensional parameters are listed in Table 1 as well.

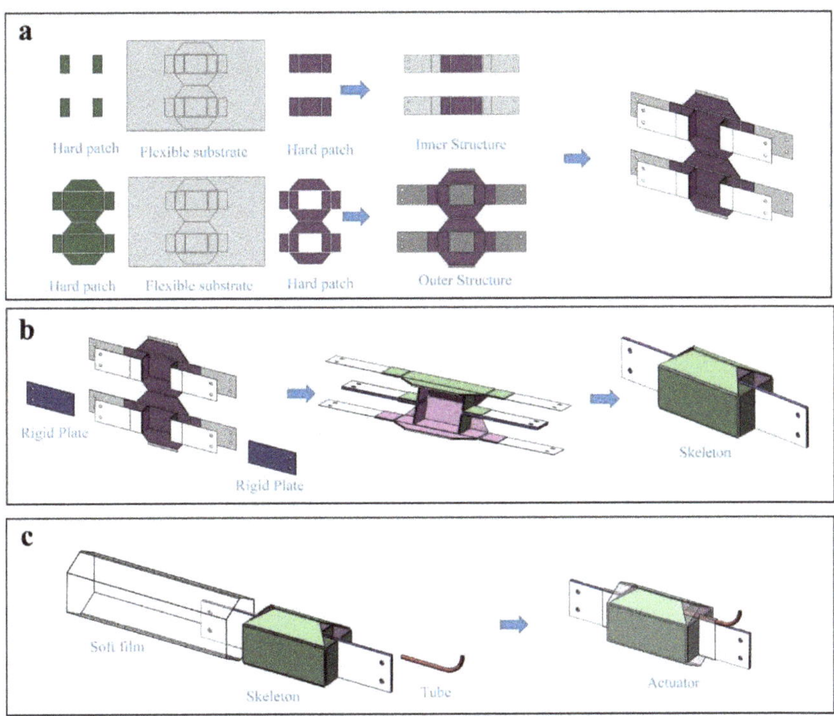

Figure 4. The fabrication process of the actuator with a patched skeleton: (**a**) Paste the hard patch on the flexible substrate; (**b**) Combine the rigid patch and the soft substrate into a skeleton; (**c**) Seal the skeleton with a flexible film to obtain an actuator.

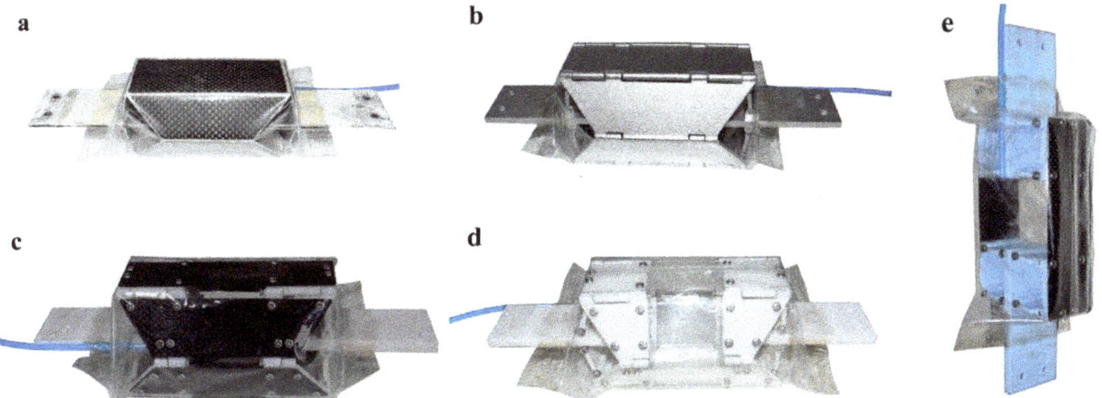

Figure 5. Actuators: (**a**) Actuator with a patched skeleton; (**b**) Actuator with a machined skeleton; (**c**) CFP-beamed actuator; (**d**) PET-beamed actuator; (**e**) Actuator without side support.

2.3.3. 3D Printed Actuator

3D printing technology can significantly promote the development of soft actuators [1]. We adopt 3D printing to make the skeleton to reduce the fabrication cost. This skeleton is assembled from 3D-printed parts with elastic beams. The size and stiffness of the skeleton are adjustable by altering the size and material of the elastic beam. It results in actuators of different performances.

First, we use a 3D printer to print the skeleton plates and a laser cutting machine to cut carbon fiber plates and PET sheets to obtain elastic beams. Second, we assemble the resultant parts to make up three skeletons and seal each with a 0.1 mm thick TPU film to form three actuators, as shown in Figure 5. The weight of the actuators with a CFP beam and a PET beam is 242 g and 240 g, respectively. The sizes of the actuators with a 3D-printed skeleton are the same as those with a machined skeleton. Finally, to reflect the influence of the side support on the actuator's performance, we fabricate an actuator without side supporting structures, as shown in Figure 5e. Its size is the same as the actuator with side support.

The working principle diagram in Figure 1 only shows the actuator's main motion structure. However, it is not enough for an actuator. To fabricate an effective actuator, we must add a side support structure to back the main structure, as shown in Figure 5.

3. Results and Discussion

3.1. Experimental Setup

We designed a test platform to test the flexible actuators and evaluate their performance, as shown in Figure 6. One end of an actuator is connected to a tension-compression sensor to measure an actuator's output force. Another end links to an electric push rod to control the actuator's stretching. Moreover, a displacement sensor is mounted to measure the actuator's displacement. The pressure of an actuator is derived from a vacuum pump with an air-vacuum proportional valve and a solenoid valve.

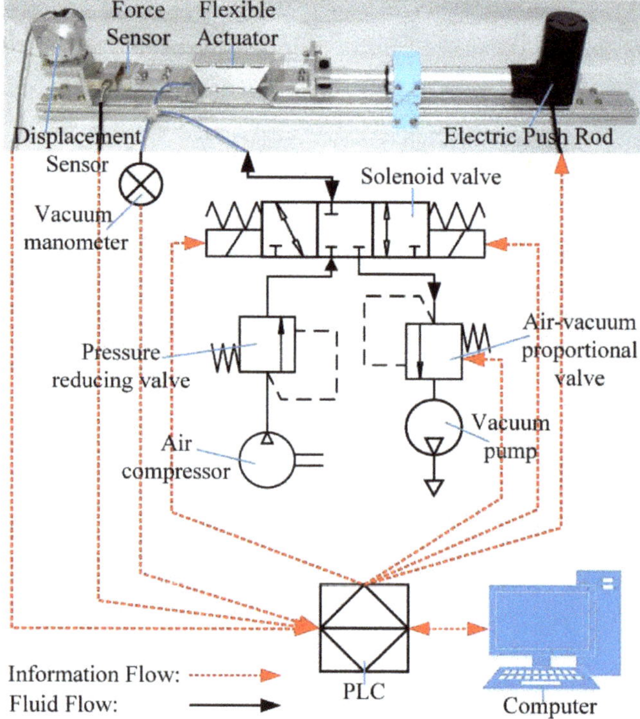

Figure 6. Test platform.

The main components of the test platform include a pull rope displacement sensor MPS-S-1000 mm from Shenzhen Milont Technology Co., LTD, Shenzhen, China, a tension-compression sensor LLBLS-I from Shanghai Longlvdianzi Technology Co., Ltd., Shanghai, China, a vacuum manometer ZSE30AF-01-E, an air-vacuum proportional valve ITV0090-3BL, solenoid valve VEX3121-025DZ-FN, and pressure reducing valve AW20-01BG-A from SMC, Tokyo, Japan, a vacuum pump 550D, from Taizhou Fujiwara Tools Co., Ltd, Taizhou, China.

3.2. Performance Test

3.2.1. Test Results and Analysis

The test results of the actuators are shown in Figures 7–11, and the test statistics are listed in Tables 2–5, respectively. To evaluate the performance of the models, Mean Absolute Error (MAE) and Mean Absolute Percentage Error (MAPE) are used as statistical means.

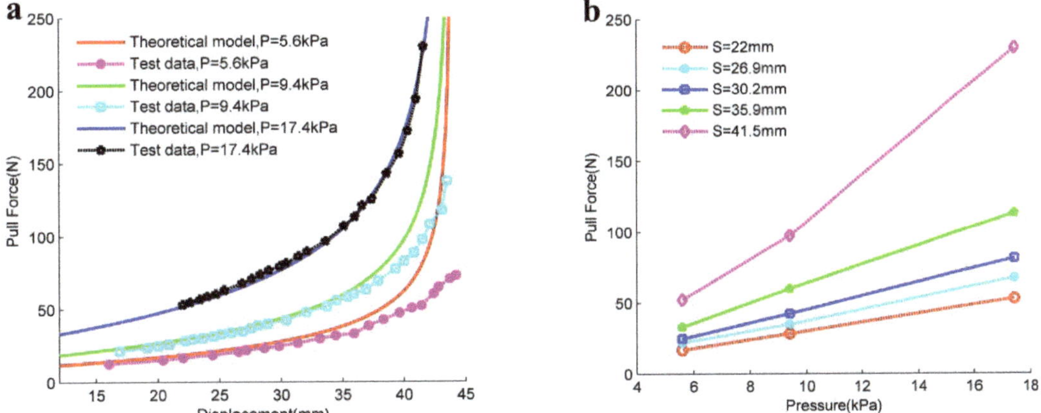

Figure 7. Test results of the actuator with a patched skeleton: (**a**) Force-displacement data; (**b**) Force-pressure data.

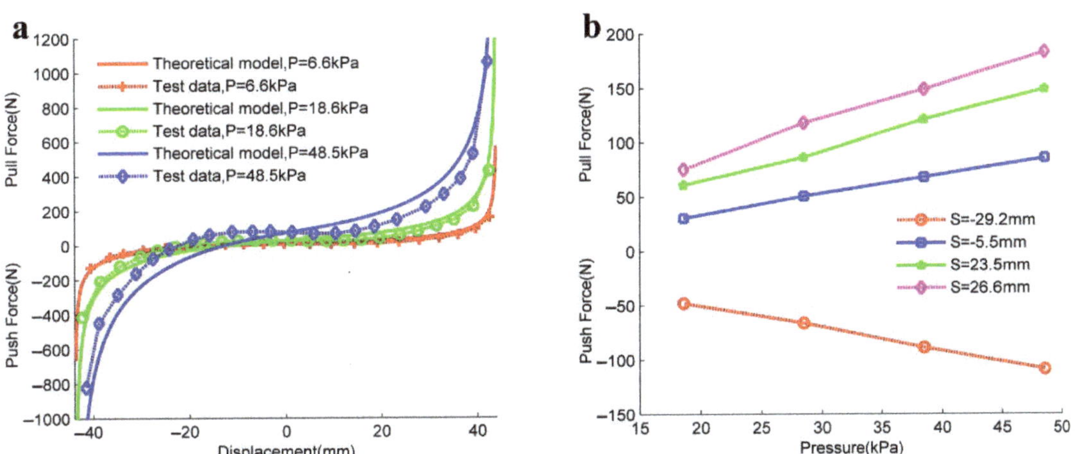

Figure 8. Test results of the actuator with a machined skeleton: (**a**) Force-displacement data; (**b**) Force-pressure data.

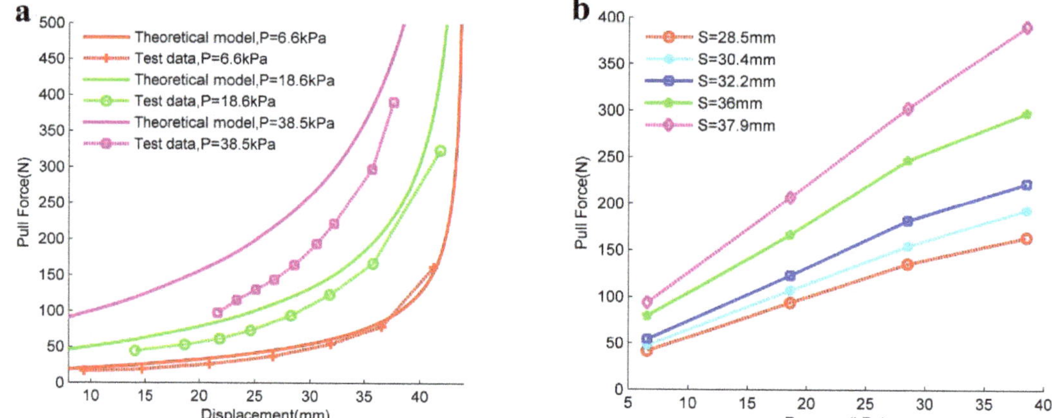

Figure 9. Test results of the PET-beamed actuator with a 3D-printed skeleton: (**a**) Force-displacement data; (**b**) Force-pressure data.

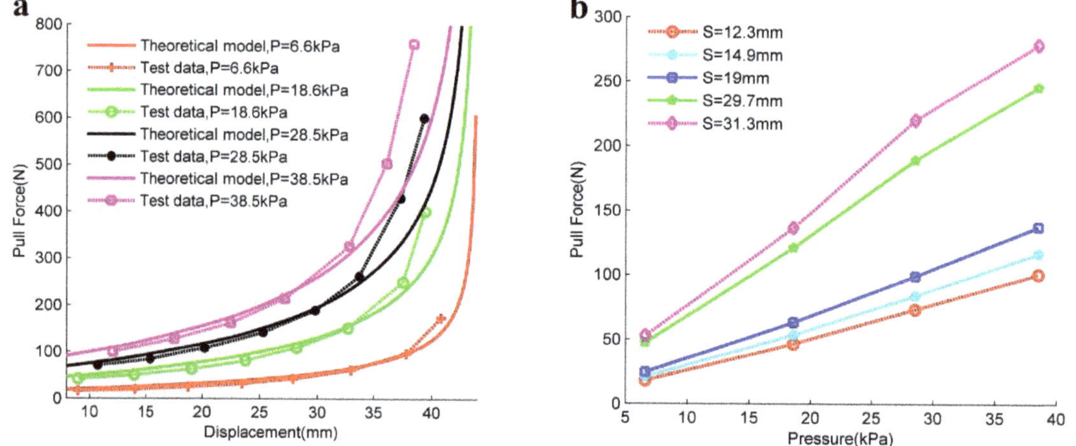

Figure 10. Test results of the CFP-beamed actuator with a 3D-printed skeleton: (**a**) Force-displacement data; (**b**) Force-pressure data.

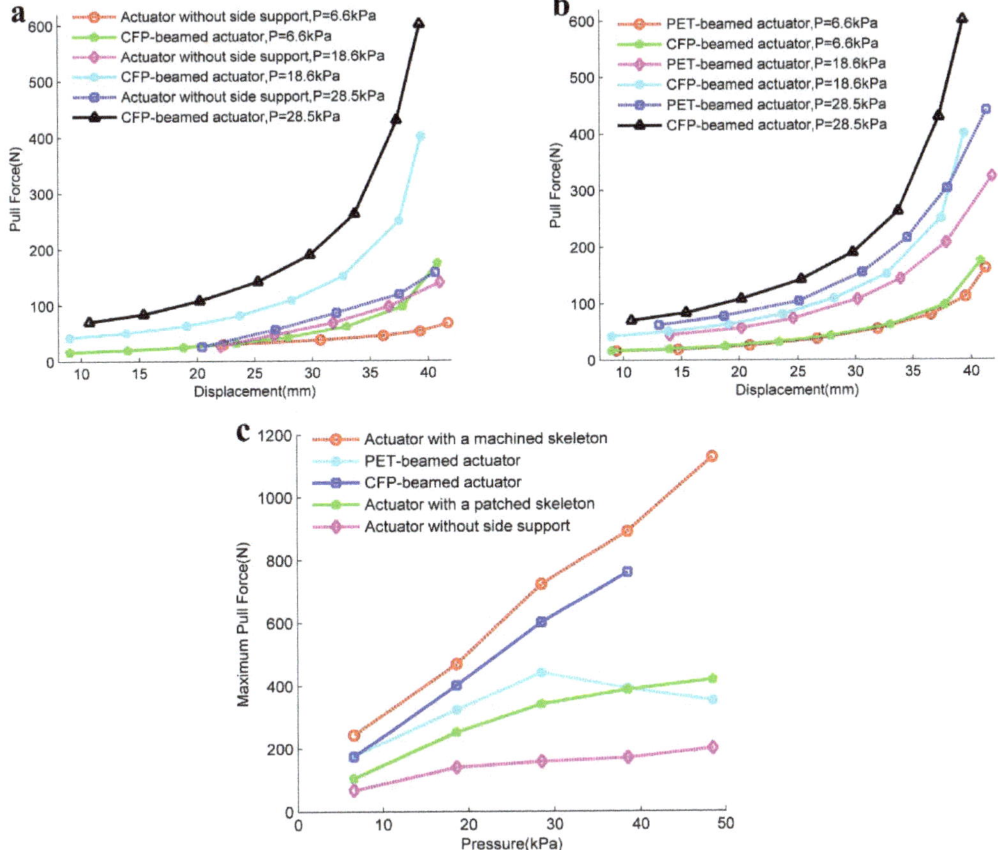

Figure 11. Test results of the actuators: (**a**) Comparison between the actuator without side supporting structure and the CFP-beamed actuator with a 3D-printed skeleton; (**b**) Comparison between PET-beamed actuator and CFP-beamed actuator; and (**c**) Comparison of the maximum output force of different actuators.

Table 2. Summary of the test data of the actuator with a patched skeleton.

Test#	Pressure (kPa)	Peak Force (N)	MAE	MAPE	Maximum Absolute Error
1	5.6	72.9	32.84	79.33	309.96
2	9.4	137.9	14.55	35.77	154.84
3	17.4	229.8	2.60	3.19	7.69

Table 3. Summary of the test data of the actuator with a machined skeleton.

Test#	Pressure (kPa)	Maximum Tension Force (N)	Maximum Thrust Force (N)	MAE	MAPE	Maximum Absolute Error
1	6.6	243.3	−125.2	11.82	13.82	42.86
2	18.6	468.1	−410.1	30.58	37.97	168.86
3	28.5	722.5	−409.5	46.50	52.37	144.47
4	38.5	889.3	−628.9	67.85	77.14	199.03
5	48.5	1125.9	−818.2	86.96	100.01	244.58

Table 4. Summary of the test data of the PET-beamed actuator.

Test#	Pressure (kPa)	Peak Force (N)	MAE	MAPE	Maximum Absolute Error
1	6.6	177	6.81	6.91	8.41
2	18.6	324.4	29.89	36.14	104.76
3	28.5	441	47.40	52.93	128.42
4	38.5	390.7	73.34	73.59	86.26
5	48.5	352.5	117.67	117.95	130.56

Table 5. Summary of the test data of the CFP-beamed actuator.

Test#	Pressure (kPa)	Peak Force (N)	MAE	MAPE	Maximum Absolute Error
1	6.6	177	6.81	6.91	8.41
2	18.6	324.4	29.89	36.14	104.76
3	28.5	441	47.40	52.93	128.42
4	38.5	390.7	73.34	73.59	86.26
5	48.5	352.5	117.67	117.95	130.56

As shown in Figures 7a, 8a and 9a, the output force of the actuator with a patched skeleton, the actuator with a machined skeleton, and the PET-beamed actuator with a 3D-printed skeleton is slightly less than the force predicted by their theoretical model. The reason may be that the soft films are trapped into the skeleton due to pressure difference and offsets that are part of the output force of the skeletons. As shown in Figure 10a, the output force of the CFP-beamed actuator with a 3D-printed skeleton is slightly less than the predicted value obtained by the theoretical model in the first half of the moving process. It is mainly affected by the soft film. The output force of the second half of the moving process is slightly greater than the theoretical model's predicted value due to the slight bending of the elastic beam to the inside.

As shown in Figures 7a, 8a, 9a and 10a, the overall shapes of the output force curves of each actuator are almost consistent with their theoretical curves. Under constant pressure, the output force of the actuators increases gradually with the arising displacement, and the growth is faster and faster. By the experimental results, as the pressure difference increases, the deviation between the output force and the model-predicted value gradually increases with the film shrinking towards the skeleton inside. In addition, we observed an approximately linear relationship between the output force of the actuator and the pressure at a given displacement, as shown in Figures 7b, 8b, 9b and 10b. Again, it is consistent with the prediction of the theoretical model established in Section 2.2.

The flexible substrate between two adjacent hard patches acts as a rotating shaft in the patched skeleton. Nevertheless, the rotating shaft can remain stable only when the flexible substrate is under tension. Therefore, the actuator with a patched skeleton can only work effectively in contraction. In the experiment, the actuator can produce a pull force of 450 N under pressure 45 kPa, equivalent to 600 times the actuator's gravity. Meanwhile, the skeleton has undergone a large deformation, and continuing to increase the pressure leads to damage to the actuator.

The actuator with a machined skeleton can bear the force in all directions and work during elongation and contraction. When the pressure is 48.5 kPa, the actuator can generate a pull force of 1125.9 N, 248.5 times its gravity, and a thrust force of 818.2 N, 180.6 times its gravity, as given in Table 3. Considering the experimental safety, we do not further test whether its mechanical properties provided higher pressure. Equation (24) and Figures 7b, 8b, 9b and 10b show an approximately linear relationship between the actuator's output force and the given pressure. The maximum vacuum pressure in the air can reach about -101kPa. Therefore, the tests can only examine about half of the potential of the actuator.

The actuators with a 3D-printed skeleton can transmit the force through the elastic beam. Therefore, its mechanical performances depend on the stiffness of the elastic beam to a certain extent. The stiffness of the PET beam is relatively minor, making it challenging to bear push force in the length direction. Thus, the actuator with a PET-beamed 3D-printed skeleton can produce a greater pull force but a relatively smaller push force. By testing, the PET-beamed actuator can reach a maximum pull force of 595.6 N, 253.4 times its gravity, but the full thrust generated is only 79N. In contrast, although the stiffness of the CFP beam is relatively higher, the tension generated by the actuator with a CFP-beamed 3D-printed skeleton is far greater than the thrust. Experimental tests show that it can achieve a maximum pull force of 759.4 N, 320.4 times its gravity, but the full thrust generated is only 60 N.

There is a particular gap at each joint of the skeleton in all the actuators mentioned above. It leads to a certain motion redundancy. Under the pressure difference between the inner and outer sides, the actuators can stabilize the state by mutual restriction between the skeleton components. The experimental results show that increasing the skeleton's overall stiffness or the actuator's motion amplitude can improve the actuator's structural stability. In addition, as the pressure difference between the inner and outer sides increases, the structural stability of the actuator is depressed.

3.2.2. Performance Comparison among Different Actuators

To explore the effect of side support on the performance of an actuator, we compare the CFP-beamed actuator with a 3D-printed skeleton and the actuator without side support. As shown in Figure 11a, the output force of the former is significantly greater than that of the latter concerning the same displacement under a certain pressure. Thus, the side support dramatically improves the output force of an actuator. On the other hand, the side support increases the internal restriction in a skeleton and significantly enhances the morphological stability of an actuator.

Similarly, we compared the PET-beamed actuator and the CFP-beamed actuator to examine the influence of the skeleton's stiffness on the actuator's performance. Both have a 3D-printed skeleton. As shown in Figure 11b, the output force of the former actuator is significantly greater than that of the latter regarding the same displacement under a given pressure. Note that the PET beam is more apt to dent into the inner of the skeleton than the CFP beam. It significantly reduces the effective stroke of the PET-beamed actuator. Therefore, with the same displacement and given pressure, the greater the skeleton's stiffness, the greater the actuator's output force, and the higher the actuator's effective stroke.

On the other hand, actuators made of different materials have different mechanical properties. For example, as shown in Figure 11c, the greater the skeleton's stiffness, the greater the full output force of an actuator under specific pressure.

Note that the stiffness of the PET beam is small, and excessive elastic deformation occurs when the pressure is too large, which significantly reduces the effective area of external pressure and the output force. Therefore, in Figure 11c, once the pressure exceeds 30 kPa, the maximum output force of the actuator with a PET beam decreases.

3.3. Demonstration of Actuators

Our flexible actuators have some characteristics unexplored before in the current work. These characteristics include dual-mode work, high force retention, large output force, and some explosiveness. To illustrate them, we design four demonstrations, i.e., push and pull, clamping, single-arm-assisted motion, and motion in water.

Figure 12a–d illustrates two work modes of our actuators, i.e., push and pull in horizontal and vertical directions.

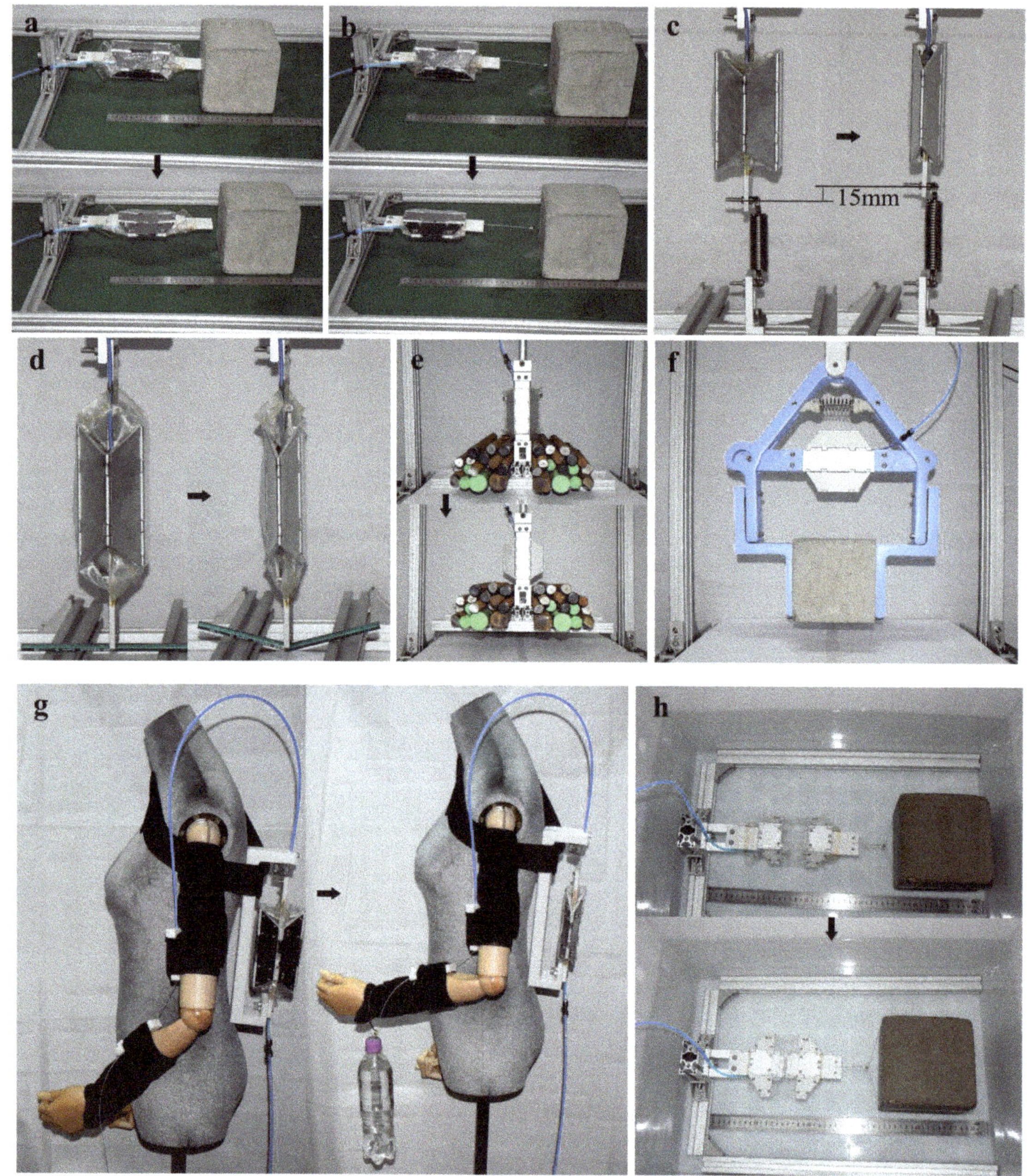

Figure 12. Demonstration of the use of our actuators: (**a**) The CFP-beamed actuator pushes and (**b**) pulls a concrete block (8.1 kg); (**c**) The actuator with a machined skeleton pulls a spring, and (**d**) punches a pencil to break; (**e**) The actuator with a machined skeleton lifts rebars at a total weight of 60.5 kg; (**f**) A clamping gripper clamps a concrete block; (**g**) The CFP-beamed actuator assists the movement of the model's arm; (**h**) The PET-beamed actuator moves in the water.

First, the CFP-beamed actuator with a 3D-printed skeleton can push and pull an 8.1 kg concrete block horizontally. In the beginning, the actuator in the push mode cannot move the concrete block away. However, once the pressure increases to 45 kPa, the actuator shoves the concrete block 4 cm away instantaneously while it stretches only 2 cm on its own, as shown in Figure 12a and Movie S1. Then, when the actuator works in the pull mode, it can pull the concrete block back 2 cm under the same pressure, as shown in Figure 12b and Movie S2.

Next, the actuator with a machined skeleton working in the pull mode can stretch a spring with an elastic coefficient of 3 N/mm for 15 mm under the pressure of 12 kPa, as shown in Figure 12c and Movie S3. Besides, the actuator in push mode can break a pencil under the pressure of 25 kPa, as shown in Figure 12d and Movie S4.

Our actuator with a machined skeleton can produce a vast output force. As demonstrated in Figure 12e and Movie S5, the actuator lifts the rebars of 60.5 kg instantly for about 7 mm without assistance when the driving pressure is 60 kPa. The characteristics of the actuator, which can quickly generate a considerable output force in a short time, are similar to the explosive force of biological muscles. It confirms again that our actuators have some explosive force like human muscles [60,61]. The actuator's lifting status remains unchanged by keeping the same pressure. It shows that the actuator has excellent force retention.

We designed a 3D-printed clamping jaw driven by the actuator with a machined skeleton. This clamping jaw can clamp the concrete block and keep the clamping when an electric push rod lifts it and the block for 22 s, as shown in Figure 12f and Movie S6. In the above process, the clamping still holds with no signs of loosening. It shows our actuators' high force retention again.

Additionally, we designed a power-assisted structure for a single arm using the CFP-beamed actuator with a 3D-printed skeleton, as shown in Figure 12g and Movie S7. The actuator tows a wood model's forearm with a Bowden cable. First, the structure bends the forearm 30° from the original position. Next, we hang a 500 mL bottle of water on the wrist when the forearm turns nearly to the horizontal position. Due to the actuator's force retention, the forearm always keeps the status as long as the actuator works.

Our actuators can be driven by liquid under negative pressure as well. For example, we use a large syringe with water to drive the PET-beamed actuator with a 3D-printed skeleton. The hydraulic actuator can pull the concrete block to slide 3 cm in water, as shown in Figure 12h and Movie S8.

4. Conclusions

We propose a novel flexible actuator structure coupling hardness with softness and fabricate five prototypes with different materials. The performance test shows that our actuators driven by negative pressure fluid can simultaneously output large force and achieve high force retention. Additionally, it can either stretch or shrink and produce human-muscle-like explosive force. The five actuators' performances vary. The actuator with a machined skeleton can reach a full pull force of 1125.9 N and a maximum thrust force of 818.2 N. The actuator with a patched skeleton can output the maximal force 600 times its weight. We demonstrated a series of applications of our actuators, including stretching, shrinking, clamping, power assistance, and underwater movement.

Further, our actuators can be used for multiple purposes due to their excellent properties and performances.

- They can work in air or water, driven by negative pressure fluid, air, or liquid. In addition, they are portable thanks to their folding structures.
- They are suitable for pushing or pulling heavy loads, owing to their large output force in a small range of strokes in two ways.
- In some scenes requiring highly reliable clamping, they can drive a robotic end effector to clamp a heavy object with high force retention. On the other hand, they can occasionally adapt well to the gradually increasing load.
- They can also apply to scenes where explosive force is required.

- Moreover, our actuators can also be combined in series or parallel to expand their applications.

On the other hand, our actuator also has some limitations. First, the actuator's work mode is determined by its initial state; thus, it cannot switch the work mode automatically. Second, the actuator's motion range is limited by the size of its skeleton's horizontal and vertical plates. Our ongoing work is to improve our actuator and further design an actuator for bending motion.

Supplementary Materials: The following supporting information can be downloaded at: https://www.mdpi.com/article/10.3390/act12010035/s1, Movie S1: The CFP-beamed actuator pushes a concrete block; Movie S2: The CFP-beamed actuator pulls a concrete block; Movie S3: The actuator with a machined skeleton pulls a spring; Movie S4: The actuator with a machined skeleton punches a pencil; Movie S5: The actuator with a machined skeleton lifts rebars; Movie S6: A clamping gripper clamps a concrete block; Movie S7: The CFP-beamed actuator assists the model's arm; Movie S8: The PET-beamed actuator moves in the water.

Author Contributions: Conceptualization, Z.S., W.Z. and J.L.; methodology, Z.S., W.Z., J.L. and I.-M.C.; validation, Z.S., W.Z. and Z.Z.; writing—original draft preparation, Z.S., W.Z. and Z.Z.; writing—review and editing, J.L., I.-M.C. and Z.S. All authors have read and agreed to the published version of the manuscript.

Funding: This research was funded in part by the National Key R&D Program of China (Grant No. 2021YFF0500900), National Natural Science Foundation of China (No. 61773115), and Shenzhen Fundamental Research Program (No. CYJ20190813152401690).

Institutional Review Board Statement: Not applicable.

Informed Consent Statement: Not applicable.

Data Availability Statement: Not applicable.

Conflicts of Interest: The authors declare no conflict of interest.

References

1. Stano, G.; Percoco, G. Additive manufacturing aimed to soft robots fabrication: A review. *Extreme Mech. Lett.* **2021**, *42*, 101079. [CrossRef]
2. Zaidi, S.; Maselli, M.; Laschi, C.; Cianchetti, M. Actuation Technologies for Soft Robot Grippers and Manipulators: A Review. *Curr. Robot. Rep.* **2021**, *2*, 355–369. [CrossRef]
3. Pan, M.; Yuan, C.; Liang, X.; Dong, T.; Liu, T.; Zhang, J.; Zou, J.; Yang, H.; Bowen, C. Soft Actuators and Robotic Devices for Rehabilitation and Assistance. *Adv. Intell. Syst.* **2022**, *4*, 2100140. [CrossRef]
4. Calisti, M.; Giorelli, M.; Levy, G.; Mazzolai, B.; Hochner, B.; Laschi, C.; Dario, P. An octopus-bioinspired solution to movement and manipulation for soft robots. *Bioinspir. Biomim.* **2011**, *6*, 036002. [CrossRef] [PubMed]
5. Shahid, Z.; Glatman, A.L.; Ryu, S.C. Design of a Soft Composite Finger with Adjustable Joint Stiffness. *Soft Robot.* **2019**, *6*, 722–732. [CrossRef] [PubMed]
6. Lin, N.; Wu, P.; Wang, M.; Wei, J.; Yang, F.; Xu, S.; Ye, Z.; Chen, X. IMU-Based Active Safe Control of a Variable Stiffness Soft Actuator. *IEEE Robot. Autom. Lett.* **2019**, *4*, 1247–1254. [CrossRef]
7. Li, Y.; Chen, Y.; Ren, T.; Li, Y.; Choi, S.H. Precharged Pneumatic Soft Actuators and Their Applications to Untethered Soft Robots. *Soft Robot.* **2018**, *5*, 567–575. [CrossRef]
8. Mazzolai, B.; Margheri, L.; Cianchetti, M.; Dario, P.; Laschi, C. Soft-robotic arm inspired by the octopus: II. From artificial requirements to innovative technological solutions. *Bioinspir. Biomim.* **2012**, *7*, 025005. [CrossRef]
9. Cianchetti, M.; Calisti, M.; Margheri, L.; Kuba, M.; Laschi, C. Bioinspired locomotion and grasping in water: The soft eight-arm OCTOPUS robot. *Bioinspir. Biomim.* **2015**, *10*, 035003. [CrossRef]
10. Yin, H.; Tian, L.; Yang, G. Design of fibre array muscle for soft finger with variable stiffness based on nylon and shape memory alloy. *Adv. Robot.* **2020**, *34*, 599–609. [CrossRef]
11. Huang, X.; Ford, M.; Patterson, Z.J.; Zarepoor, M.; Pan, C.; Majidi, C. Shape memory materials for electrically-powered soft machines. *J. Mater. Chem. B* **2020**, *8*, 4539–4551. [CrossRef] [PubMed]
12. Behl, M.; Kratz, K.; Zotzmann, J.; Nöchel, U.; Lendlein, A. Reversible Bidirectional Shape-Memory Polymers. *Adv. Mater.* **2013**, *25*, 4466–4469. [CrossRef] [PubMed]
13. Youn, J.-H.; Jeong, S.M.; Hwang, G.; Kim, H.; Hyeon, K.; Park, J.; Kyung, K.-U. Dielectric Elastomer Actuator for Soft Robotics Applications and Challenges. *Appl. Sci.* **2020**, *10*, 640. [CrossRef]

14. Moss, A.; Krieg, M.; Mohseni, K. Modeling and Characterizing a Fiber-Reinforced Dielectric Elastomer Tension Actuator. *IEEE Robot. Autom. Lett.* **2021**, *6*, 1264–1271. [CrossRef]
15. Li, W.-B.; Zhang, W.-M.; Zou, H.-X.; Peng, Z.-K.; Meng, G. A Fast Rolling Soft Robot Driven by Dielectric Elastomer. *IEEE/ASME Trans. Mechatron.* **2018**, *23*, 1630–1640. [CrossRef]
16. Liang, W.; Liu, H.; Wang, K.; Qian, Z.; Ren, L.; Ren, L. Comparative study of robotic artificial actuators and biological muscle. *Adv. Mech. Eng.* **2020**, *12*, 1687814020933409. [CrossRef]
17. Wang, H.; Chen, J.; Lau, H.Y.K.; Ren, H. Motion Planning Based on Learning from Demonstration for Multiple-Segment Flexible Soft Robots Actuated by Electroactive Polymers. *IEEE Robot. Autom. Lett.* **2016**, *1*, 391–398. [CrossRef]
18. Ishiki, A.; Nabae, H.; Kodaira, A.; Suzumori, K. PF-IPMC: Paper/Fabric Assisted IPMC Actuators for 3D Crafts. *IEEE Robot. Autom. Lett.* **2020**, *5*, 4035–4041. [CrossRef]
19. Yi, X.; Chakarvarthy, A.; Chen, Z. Cooperative Collision Avoidance Control of Servo/IPMC Driven Robotic Fish with Back-Relaxation Effect. *IEEE Robot. Autom. Lett.* **2021**, *6*, 1816–1823. [CrossRef]
20. Zhu, Z.; Bian, C.; Ru, J.; Bai, W.; Chen, H. Rapid deformation of IPMC under a high electrical pulse stimulus inspired by action potential. *Smart Mater. Struct.* **2019**, *28*, 01LT01. [CrossRef]
21. Tang, X.; Li, H.; Ma, T.; Yang, Y.; Luo, J.; Wang, H.; Jiang, P. A Review of Soft Actuator Motion: Actuation, Design, Manufacturing and Applications. *Actuators* **2022**, *11*, 331. [CrossRef]
22. Rich, S.I.; Wood, R.J.; Majidi, C. Untethered soft robotics. *Nat. Electron.* **2018**, *1*, 102–112. [CrossRef]
23. Hu, W.; Lum, G.Z.; Mastrangeli, M.; Sitti, M. Small-scale soft-bodied robot with multimodal locomotion. *Nature* **2018**, *554*, 81–85. [CrossRef] [PubMed]
24. Kim, J.; Chung, S.E.; Choi, S.-E.; Lee, H.; Kim, J.; Kwon, S. Programming magnetic anisotropy in polymeric microactuators. *Nat. Mater.* **2011**, *10*, 747–752. [CrossRef] [PubMed]
25. Kim, Y.; Yuk, H.; Zhao, R.; Chester, S.A.; Zhao, X. Printing ferromagnetic domains for untethered fast-transforming soft materials. *Nature* **2018**, *558*, 274–279. [CrossRef]
26. Kim, Y.; Parada, G.A.; Liu, S.; Zhao, X. Ferromagnetic soft continuum robots. *Sci. Robot.* **2019**, *4*, eaax7329. [CrossRef]
27. Sudhawiyangkul, T.; Yoshida, K.; Eom, S.I.; Kim, J.W. A Multi-DOF Soft Microactuator Integrated with Flexible Electro-Rheological Microvalves Using an Alternating Pressure Source. *Smart Mater. Struct.* **2021**, *30*, 085006. [CrossRef]
28. Walker, J.; Zidek, T.; Harbel, C.; Yoon, S.; Strickland, F.S.; Kumar, S.; Shin, M. Soft Robotics: A Review of Recent Developments of Pneumatic Soft Actuators. *Actuators* **2020**, *9*, 3. [CrossRef]
29. Kalita, B.; Leonessa, A.; Dwivedy, S.K. A Review on the Development of Pneumatic Artificial Muscle Actuators: Force Model and Application. *Actuators* **2022**, *11*, 288. [CrossRef]
30. Jiao, Z.; Zhuang, Z.; Cheng, Y.; Deng, X.; Sun, C.; Yu, Y.; Li, F. Lightweight Dual-Mode Soft Actuator Fabricated from Bellows and Foam Material. *Actuators* **2022**, *11*, 245. [CrossRef]
31. Hao, Y.; Gong, Z.; Xie, Z.; Guan, S.; Yang, X.; Wang, T.; Wen, L. A Soft Bionic Gripper with Variable Effective Length. *J. Bionic Eng.* **2018**, *15*, 220–235. [CrossRef]
32. Yan, J.; Xu, Z.; Shi, P.; Zhao, J. A Human-Inspired Soft Finger with Dual-Mode Morphing Enabled by Variable Stiffness Mechanism. *Soft Robot.* **2021**, *9*, 399–411. [CrossRef]
33. Sparrman, B.; du Pasquier, C.; Thomsen, C.; Darbari, S.; Rustom, R.; Laucks, J.; Shea, K.; Tibbits, S. Printed silicone pneumatic actuators for soft robotics. *Addit. Manuf.* **2021**, *40*, 101860. [CrossRef]
34. Pan, M.; Yuan, C.; Anpalagan, H.; Plummer, A.; Zou, J.; Zhang, J.; Bowen, C. Soft Controllable Carbon Fibre-Based Piezore-sistive Self-Sensing Actuators. *Actuators* **2020**, *9*, 79. [CrossRef]
35. Bell, M.A.; Gorissen, B.; Bertoldi, K.; Weaver, J.C.; Wood, R.J. A Modular and Self-Contained Fluidic Engine for Soft Actuators. *Adv. Intell. Syst.* **2021**, *4*, 2100094. [CrossRef]
36. Wang, B.; Guo, W.; Feng, S.; Hongdong, Y.; Wan, F.; Song, C. Volumetrically Enhanced Soft Actuator with Proprioceptive Sensing. *IEEE Robot. Autom. Lett.* **2021**, *6*, 5284–5291. [CrossRef]
37. Ke, X.; Jang, J.; Chai, Z.; Yong, H.; Zhu, J.; Chen, H.; Guo, C.F.; Ding, H.; Wu, Z. Stiffness Preprogrammable Soft Bending Pneumatic Actuators for High-Efficient, Conformal Operation. *Soft Robot.* **2021**, *9*, 613–624. [CrossRef] [PubMed]
38. Wirekoh, J.; Park, Y.-L. Design of flat pneumatic artificial muscles. *Smart Mater. Struct.* **2017**, *26*, 035009. [CrossRef]
39. Wirekoh, J.; Valle, L.; Pol, N.; Park, Y.-L. Sensorized, Flat, Pneumatic Artificial Muscle Embedded with Biomimetic Microfluidic Sensors for Proprioceptive Feedback. *Soft Robot.* **2019**, *6*, 768–777. [CrossRef]
40. Hu, L.; Gau, D.; Nixon, J.; Klein, M.; Fan, Y.; Menary, G.; Roche, E.T. Precurved, Fiber-Reinforced Actuators Enable Pneumatically Efficient Replication of Complex Biological Motions. *Soft Robot.* **2021**, *9*, 293–308. [CrossRef]
41. Yang, H.D.; Greczek, B.T.; Asbeck, A.T. Modeling and Analysis of a High-Displacement Pneumatic Artificial Muscle with Integrated Sensing. *Front. Robot. AI* **2019**, *5*, 136. [CrossRef]
42. Shaheen, R.; Doumit, M.; Helal, A. Design and characterization of a hyperelastic tubular soft composite. *J. Mech. Behav. Biomed. Mater.* **2017**, *75*, 228–235. [CrossRef] [PubMed]
43. Al-Fahaam, H.; Davis, S.; Nefti-Meziani, S. The design and mathematical modelling of novel extensor bending pneumatic artificial muscles (EBPAMs) for soft exoskeletons. *Robot. Auton. Syst.* **2018**, *99*, 63–74. [CrossRef]
44. Guan, Q.; Sun, J.; Liu, Y.; Wereley, N.M.; Leng, J. Novel Bending and Helical Extensile/Contractile Pneumatic Artificial Muscles Inspired by Elephant Trunk. *Soft Robot.* **2020**, *7*, 597–614. [CrossRef] [PubMed]

45. Zhu, M.; Do, T.N.; Hawkes, E.; Visell, Y. Fluidic Fabric Muscle Sheets for Wearable and Soft Robotics. *Soft Robot.* **2020**, *7*, 179–197. [CrossRef]
46. Kim, W.; Park, H.; Kim, J. Compact Flat Fabric Pneumatic Artificial Muscle (ffPAM) for Soft Wearable Robotic Devices. *IEEE Robot. Autom. Lett.* **2021**, *6*, 2603–2610. [CrossRef]
47. Naclerio, N.D.; Hawkes, E.W. Simple, Low-Hysteresis, Foldable, Fabric Pneumatic Artificial Muscle. *IEEE Robot. Autom. Lett.* **2020**, *5*, 3406–3413. [CrossRef]
48. Kulasekera, A.L.; Arumathanthri, R.B.; Chathuranga, D.S.; Gopura, R.; Lalitharatne, T.D. A thin-walled vacuum actuator (ThinVAc) and the development of multi-filament actuators for soft robotic applications. *Sens. Actuators A Phys.* **2021**, *332*, 113088. [CrossRef]
49. Yu, B.; Yang, J.; Du, R.; Zhong, Y. A Versatile Pneumatic Actuator Based on Scissor Mechanisms: Design, Modeling, and Experiments. *IEEE Robot. Autom. Lett.* **2021**, *6*, 1288–1295. [CrossRef]
50. Li, S.; Vogt, D.M.; Rus, D.; Wood, R.J. Fluid-driven origami-inspired artificial muscles. *Proc. Natl. Acad. Sci. USA* **2017**, *114*, 13132–13137. [CrossRef]
51. Li, S.; Vogt, D.M.; Bartlett, N.W.; Rus, D.; Wood, R.J. Tension Pistons: Amplifying Piston Force Using Fluid-Induced Tension in Flexible Materials. *Adv. Funct. Mater.* **2019**, *29*, 1901419. [CrossRef]
52. Lee, J.-G.; Rodrigue, H. Origami-Based Vacuum Pneumatic Artificial Muscles with Large Contraction Ratios. *Soft Robot.* **2019**, *6*, 109–117. [CrossRef] [PubMed]
53. Liu, T.; Xia, H.; Lee, D.-Y.; Firouzeh, A.; Park, Y.-L.; Cho, K.-J. A Positive Pressure Jamming Based Variable Stiffness Structure and its Application on Wearable Robots. *IEEE Robot. Autom. Lett.* **2021**, *6*, 8078–8085. [CrossRef]
54. Ranzani, T.; Gerboni, G.; Cianchetti, M.; Menciassi, A. A bioinspired soft manipulator for minimally invasive surgery. *Bioinspir. Biomim.* **2015**, *10*, 035008. [CrossRef] [PubMed]
55. Ranzani, T.; Cianchetti, M.; Gerboni, G.; De Falco, I.; Menciassi, A. A Soft Modular Manipulator for Minimally Invasive Sur-gery: Design and Characterization of a Single Module. *IEEE Trans. Robot.* **2016**, *32*, 187–200. [CrossRef]
56. De Falco, I.; Cianchetti, M.; Menciassi, A. A soft multi-module manipulator with variable stiffness for minimally invasive surgery. *Bioinspir. Biomim.* **2017**, *12*, 056008. [CrossRef]
57. Li, Y.; Chen, Y.; Yang, Y.; Wei, Y. Passive Particle Jamming and Its Stiffening of Soft Robotic Grippers. *IEEE Trans. Robot.* **2017**, *33*, 446–455. [CrossRef]
58. Yan, J.; Shi, P.; Xu, Z.; Zhao, J. A Wide-Range Stiffness-Tunable Soft Actuator Inspired by Deep-Sea Glass Sponges. *Soft Robot.* **2021**, *9*, 625–637. [CrossRef]
59. Wang, Y.; Yang, Z.; Zhou, H.; Zhao, C.; Barimah, B.; Li, B.; Xiang, C.; Li, L.; Gou, X.; Luo, M. Inflatable Particle-Jammed Robotic Gripper Based on Integration of Positive Pressure and Partial Filling. *Soft Robot.* **2021**, *9*, 309–323. [CrossRef]
60. Tillin, N.A.; Pain, M.T.G.; Folland, J.P. Contraction speed and type influences rapid utilisation of available muscle force: Neural and contractile mechanisms. *J. Exp. Biol.* **2018**, *221*, jeb193367. [CrossRef]
61. Del Vecchio, A.; Negro, F.; Holobar, A.; Casolo, A.; Folland, J.P.; Felici, F.; Farina, D. You Are as Fast as Your Motor Neurons: Speed of Recruitment and Maximal Discharge of Motor Neurons Determine the Maximal Rate of Force Development in Hu-mans. *J. Physiol.* **2019**, *597*, 2445–2456. [CrossRef] [PubMed]

 *actuators* 

Article

Design and Test of an Active Pneumatic Soft Wrist for Soft Grippers

Guangming Chen [1,*], Tao Lin [1], Shi Ding [2], Shuang Chen [3], Aihong Ji [1] and Gabriel Lodewijks [4]

1 Laboratory of Locomotion Bioinspiration and Intelligent Robots, College of Mechanical and Electrical Engineering, Nanjing University of Aeronautics and Astronautics, 29 Yudao Street, Nanjing 210016, China
2 College of Civil Aviation, Nanjing University of Aeronautics and Astronautics, No. 29 Jiangjun Avenue, Nanjing 210016, China
3 School of Mechanical Engineering, Xi'an Jiaotong University, No. 28 Xianning West Road, Xi'an 710049, China
4 School of Engineering, College of Engineering, Science & Environment, Callaghan Campus, University of Newcastle, University Drive, Callaghan, NSW 2308, Australia
* Correspondence: guangming2017@nuaa.edu.cn

Abstract: An active wrist can deliver both bending and twisting motions that are essential for soft grippers to perform dexterous manipulations capable of producing a wide range movements. Currently, the versions of gripper wrists are relatively heavy due to the bending and twisting motions performed by the motors. Pneumatic soft actuators can generate multiple motions with lightweight drives. This research evaluates a pneumatic soft wrist based on four parallel soft helical actuators. The kinematics models for predicting bending and twisting motions of this soft wrist are developed. Finite element method simulations are conducted to verify the functions of bending and twisting of this wrist. In addition, the active motions of the soft pneumatic wrist are experimentally demonstrated. Based on sensitivity studies of geometric parameters, a set of parameter values are identified for obtaining maximum bending and twisting angles for a bionic human wrist. Through simulation and experimental tests of the soft wrist for a soft gripper, the desired bending and twisting motions as those of a real human hand wrist are established.

Keywords: soft actuator; helical structure; pneumatic control; robotic gripper; bionic design

Citation: Chen, G.; Lin, T.; Ding, S.; Chen, S.; Ji, A.; Lodewijks, G. Design and Test of an Active Pneumatic Soft Wrist for Soft Grippers. *Actuators* **2022**, *11*, 311. https://doi.org/10.3390/act11110311

Academic Editor: Hamed Rahimi Nohooji

Received: 7 October 2022
Accepted: 24 October 2022
Published: 27 October 2022

1. Introduction

Soft robotic grippers can be used to grasp various objects and have been applied in many industrial practices [1–6]. To increase the dexterity of grippers, an active soft wrist is highly desired due to its ability to increase grasping ranges and perspectives [7,8]. Contemporarily, these movements of bending and twisting for the wrists of grippers are achieved by the forces and torques generated by a series of motors [9,10]. Although different motions are obtained, the multiple motors used in the wrist make it relatively heavy and they cause a high energy consumption. In addition, it is difficult to achieve specific bending or twisting angles via the adjustment of forces and torques by the motors [11].

Compared to the wrists of soft grippers whose motions are driven by motors, pneumatic soft actuators are convenient to generate bending motions [12–15] or twisting motions [16–19] using a lightweight driving system that is mostly composed of air pumps, tubes and control components. Thus, pneumatic soft actuators have been investigated for achieving multiple motions with lightweight drives [20,21]. For instances, Kurumaya et al. [22] designed a soft wrist using two types of fiber-reinforced soft actuators whose structural design is similar to the mechanism of a human wrist. The wrist consists of a bending module and a rotary module of fiber-reinforced elastic cylinders that can perform bending and rotation motions in correspondence to different segments. Fei et al. [23] developed a biaxial bidirectional cylinder-shaped soft actuator. The wrist can perform flexion/extension, ulnar/radii deviation, and bending in other directions by pressurizing different chambers. This wrist can increase the range orientations for grasping an object, making the soft gripper versatile to grip objects via various poses. Shen et al. [24] designed a soft wrist that consists of three fluidic actuators, a ball joint, and two 3D printed plates. The three soft actuators are placed triangularly and the top and bottom plates are connected by a ball joint. When one actuator is elongated, the other two will be contracted accordingly, which will result in the bending motion along with contracted actuators. The dexterity of the hand has been improved due to the advantage of increased motions.

Therefore, the available soft wrists can perform either bending or twisting motions at varied segments. They are not yet able to achieve both bending and twisting movements in one segment [25], which thus could not resemble twisting behaviors of a human wrist when grasping from various perspectives. To enhance grasping ranges and perspectives for robotic grippers, it is essential to develop a pneumatic active soft wrists that can present bending and twisting motions in one segment.

Fiber-reinforced pneumatic soft actuators have the advantage of performing multiple motions by a varying chamber structure, orientation of winding fibers and pressure [21,26]. The winding fibers on soft actuators also reinforce their strength with respect to high air pressure [17]. For pneumatic soft helical actuators, the characteristics of the helical structures of the soft actuators are decisive to their twisting motions [18,19]. In addition, antagonistic actions can be utilized to increase the overall stiffness of the whole structure [27]. Accordingly, with innovative structure design by using different helical actuators, the motions of bending and twisting and sufficient stiffness for an active soft wrist can be achieved for a single segment.

This paper contributes to the research field by presenting an active soft wrist that can achieve bending and twisting motions for one segment. This segment consists of four parallel soft helical actuators. This driving system is lightweight and energy-saving because only one air pump is used. A kinetic analysis, finite element simulations, and experimental tests were carried out to evaluate the wrist's bending and twisting angles for different air pressures. Moreover, the influences of geometric parameters on bending and twisting angles are predicted by using finite element method simulations. Using a set of optimized geometric parameters, a soft wrist was manufactured and the dexterity is demonstrated based on the performances on a soft pneumatic hand.

2. Structure Design

This section illustrates the structural design of the pneumatic active soft wrist. First, the structure and motions of two pneumatic actuators of opposite helical structures are introduced. On this basis, the soft wrist design composed of four helical actuators is provided, and the mechanisms for producing bending and twisting motions are explained.

2.1. Structure of Pneumatic Helical Actuators

Figure 1a presents two pneumatic soft helical actuators of the same geometric sizes but opposite helical directions. Each actuator possesses a soft chamber whose outer surface is winded by the fiber of a rectangular strip. The chamber has a universal outer and inner helical structure. Under pressurization, the winding rectangular fiber restrains the radial expansion, such that the soft actuators elongate and twist in accordance with the helical directions [26]. Due to the different orientations of the winding fiber, the actuators twist in different directions [17]. As denoted in Figure 1b, the chamber structural parameters include the overall length L, the inner hole radius r_1, the inner helix radius r_2, the inner helix width b, the helix pitch d, the outward layer thickness t_1 and the inward layer thickness t_2. Note that the width of the fiber is equal to the inner helix width b for sufficient restraining inflation of the chamber.

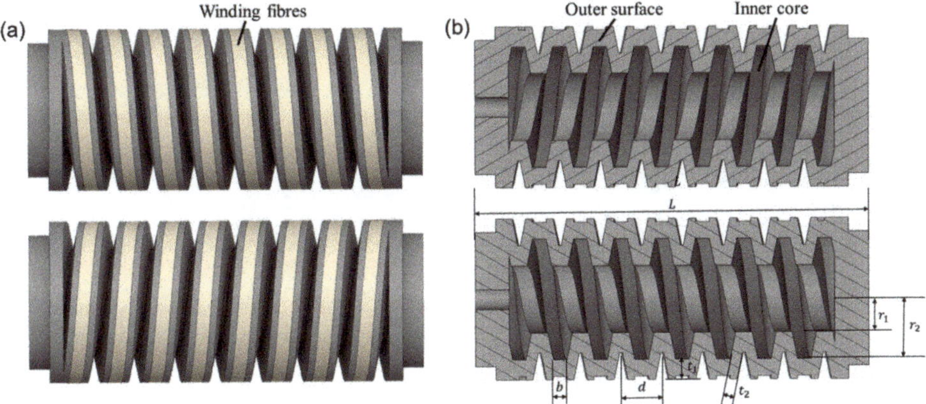

Figure 1. Pneumatic soft helical actuators: (**a**) two opposite helical structures; (**b**) longitudinal sections and the geometric parameters.

2.2. Structure Design of the Pneumatic Soft Wrist

Section 2.1 explained the elongation and twisting characteristics of a single pneumatic helical soft actuator. By bonding two soft actuators of opposite helical directions that are under pressurization (i.e., *A*&*B* of Figure 2a), the twisting effects of the two actuators are offset. When adding another two such actuators (i.e., *C*&*D* of Figure 2a), the longitudinal extension of the pressured actuators (*A*&*B*) can be restrained. In that case, the whole structure will produce a forward bending motion. Figure 2b illustrates a cross-sectional view and the helical directions of the soft wrist. With this structure, twisting of the whole structure can also be achieved by pressurization of diagonal actuators (i.e., *A*&*C* or *B*&*D*). When all of the four helical chambers are pressurized, the twisting torques of the four actuators are offset due to antagonistic actions, and the stiffness of the wrist can be improved [27].

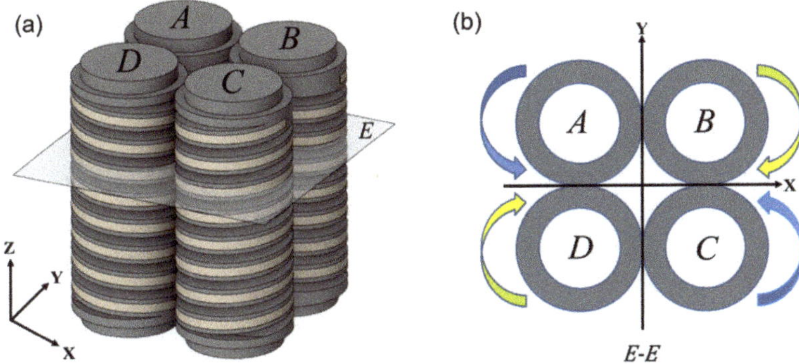

Figure 2. Design of a pneumatic soft wrist: (**a**) a soft wrist module; (**b**) cross-section of the soft wrist.

Table 1 illustrates four bending motions and two twisting motions caused by the pressurizations of different chambers. The green color of the chamber cores represent the actuators that are inflated and the green arrow denotes the motion direction of the soft wrist. For the bending motion shown in the first row of Table 1, the chambers *A* and *B* are pressurized, and the bending occurs along the negative Y direction. Likewise, the bending motions backward, left, and right can be achieved in accordance with the pressurization of neighboring chambers. The left or right twisting motions around the Z axis can be achieved similarly with the pressurization of the diagonal chambers. To increase the stiffness of the soft wrist while maintaining their motions, the same amount of air pressure can be simultaneously applied to all four chambers.

Table 1. Soft wrist motions corresponding to pressurization of different chambers.

Motions	Pressured Chambers	Directions of Motions
Bending forward (flexion)	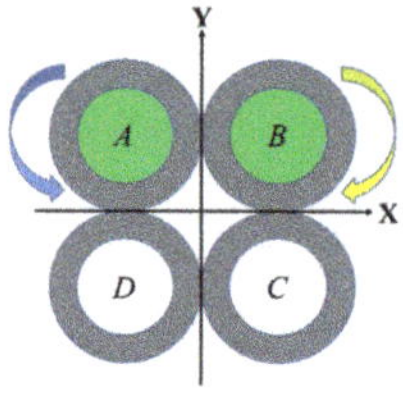 Y, X; A, B, D, C	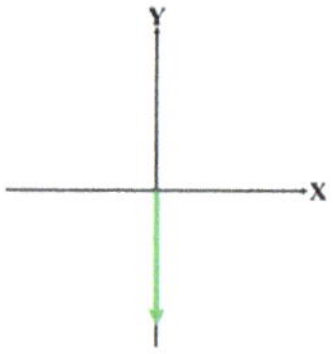 Y, X
Bending backward (extension)	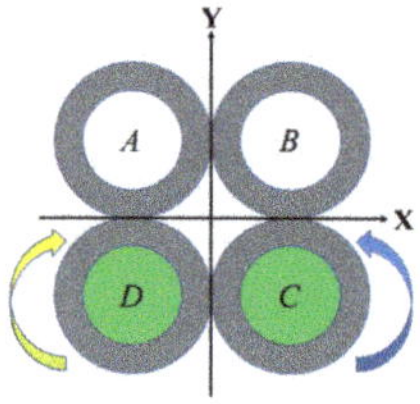 Y, X; A, B, D, C	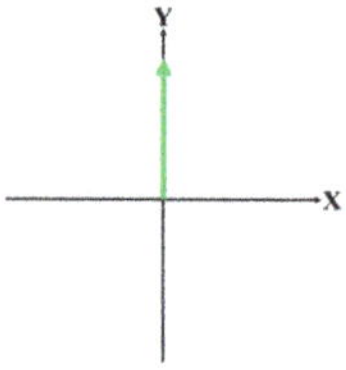 Y, X
Bending left (abduction)	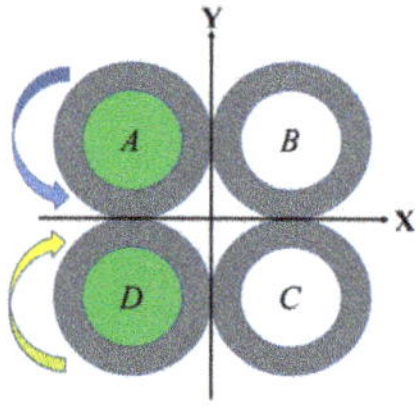 Y, X; A, B, D, C	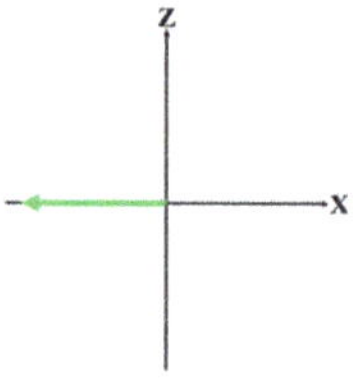 Z, X
Bending right (adduction)	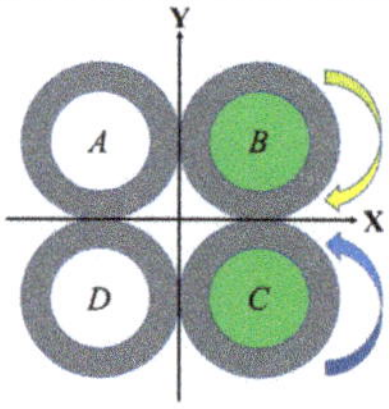 Y, X; A, B, D, C	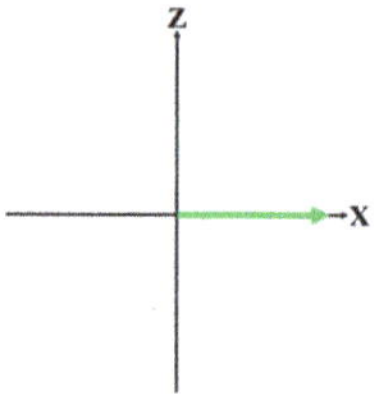 Z, X
Twisting left (supination)	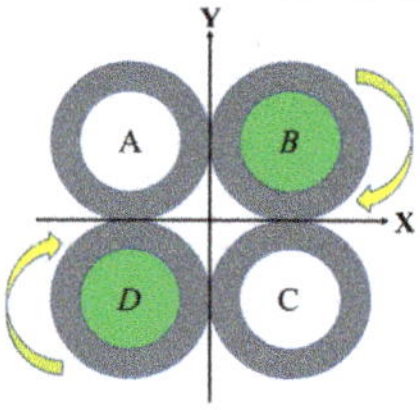 Y, X; A, B, D, C	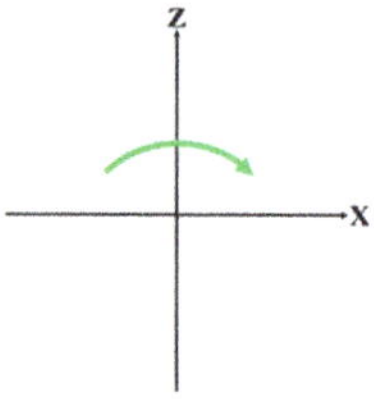 Z, X
Twisting right (pronation)	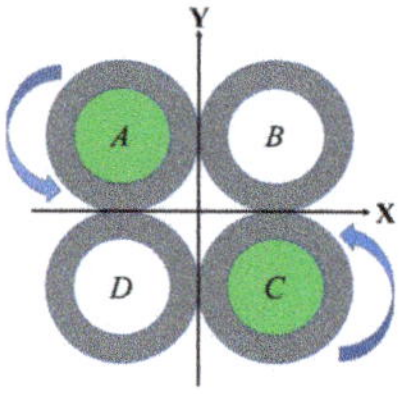 Y, X; A, B, D, C	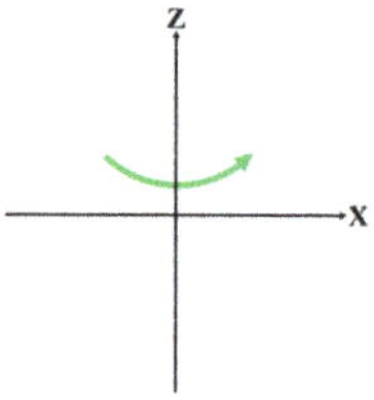 Z, X

Summarizing, this subsection presented the design of a soft joint based on four pneumatic soft helical actuators. The six motions of direction of the soft joint can be obtained by the pressurization of specific chambers. To assess the bending and twisting motions of the soft wrist, a kinematic analysis will be presented in the next section.

3. Kinematic Analysis

The bending and twisting angles can be used to assess the performances of the soft wrist under different applied air pressures [15,19]. To establish the relationships between the input pressure and the angles of bending and twisting of the soft wrist, this section presents approaches for theoretical predictions based on the kinematics analysis of a single actuator.

3.1. Bending Model

Figure 3a presents the bending model of the soft wrist subjected to two upper pressured adjacent actuators. The left side of the wrist is fixed and the bending motion of the soft wrist is triggered when the two upper actuators equally compress the two lower actuators. To deduce the bending equation for the soft wrist, the kinematic of bending for a single pressured actuator is studied. Figure 3b shows the bending forces and torques of the soft actuator of the longitudinal section. The transverse circular section of the actuator chamber bottom is shown in Figure 3c. Based on the principle that the bending moment caused by the input air pressure equates to that generated by material deformations during pressurization [28], the bending angle in respect to the applied pressure can be derived as is given below.

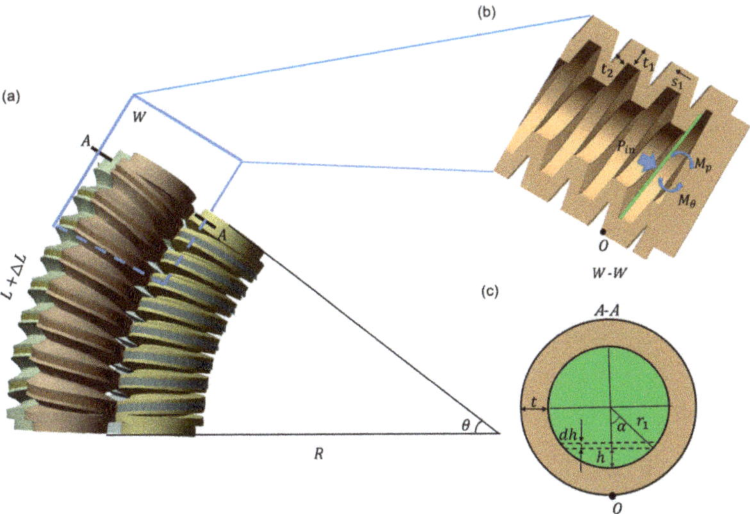

Figure 3. Bending model of the soft wrist: (**a**) bending of four actuators; (**b**) longitudinal section of a pressured actuator; (**c**) actuator chamber transverse section.

The actuator length is denoted by L, the bending angle by θ, and the curvature radius by R; M_P is the external moment imposed by the air pressure about bottom edge O; M_θ is the material moment generated by the actuators' inner wall resistance. The force df, generated by air pressure P_{in} on a micro-area of height dh, is:

$$\mathrm{d}f = P_{in}S = P_{in} \cdot 2r_1 \sin\alpha \cdot \mathrm{d}h \quad (1)$$

Using the geometric characteristic $h = r_1(1 - \cos\alpha)$, the bending moment M_P of the single actuator with respect to the bottom edge O can be expressed as:

$$M_p = \int_0^{2r_1} \mathrm{d}f \cdot (h + t) = \pi {r_1}^2 P_{in}(r_1 + t) \tag{2}$$

where t is the average equivalent thickness of the actuator chamber, which can be estimated by:

$$t = t_1 + t_2 + (r_2 - r_1) \tag{3}$$

For the helical soft chamber, the axial principle stress is the main internal stress responsible for the bending of the actuator [29]. For the circular section of the chamber, the axial principal stretch ratio λ_1 is given by:

$$\lambda_1 = \frac{t - c}{L} \cdot \frac{d}{b} \frac{180^\circ \cdot \theta}{\pi} + 1 \tag{4}$$

Then, the chamber material intrinsic resistance moment can be calculated as:

$$M_i = 2\int_0^{\pi} G_0\left(\lambda_1 - \lambda_1^{-3}\right)\left(\int_0^{t}\left((r_1 + t)(r_1 + c) - (r_1 + c)^2 \cos\alpha\right) L d\alpha\right) dc \tag{5}$$

Solving the equation of moment equilibrium $M_p = M_i$, the relationship between the input pressure of the actuator and its bending angle can be obtained.

3.2. *Twisting Model*

Figure 4a presents the twisting model of the soft wrist where two diagonal actuators are pressured whilst the bottom of the wrist is fixed. The two pressured actuators synchronously twist and cause the wrist to twist. The kinematics for a single twisting actuator are investigated to derive the twisting angle of the soft wrist. The twisting motion of the actuator is generated by the torques approximately with respect to its central axis [19]. The longitudinal section of a twisting actuator is shown in Figure 4b. The torques of the pressured actuator are given in Figure 4c. The equation of the moment balance equilibrium for the torque generated by air pressure and the torque generated by the material is used to derive the twist angle with respect to applied pressure.

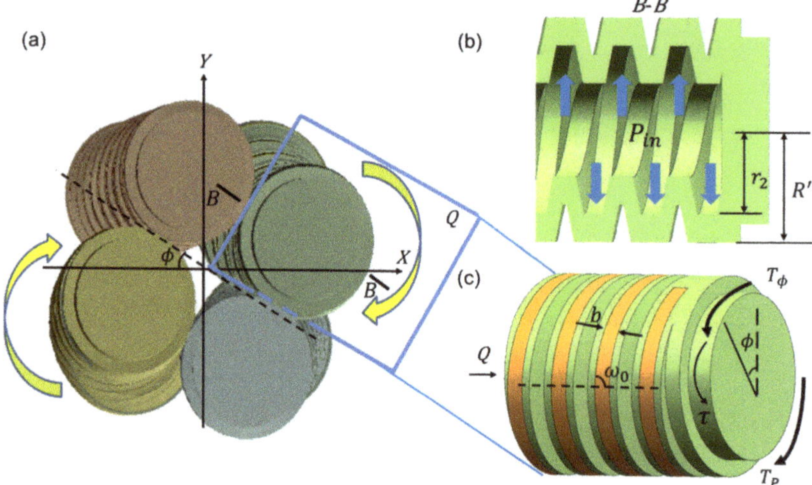

Figure 4. Twisting model of the soft wrist: (**a**) twisting of four actuators; (**b**) longitudinal section of a pressured actuator; (**c**) shear model of a pressured actuator.

The twisting of the actuator is mainly caused by the pressure applied on the inner helical structure surface of the chamber [18]. The parametric equation for the helix curvature of the inner helical structure can be expressed as:

$$\begin{cases} x = R' \cos(\omega) \\ y = R' \sin(\omega) \\ z = k\omega \end{cases} \tag{6}$$

where ω is the helix angle; $k = d/2\pi$. Neglecting the mutual resistance of the actuators, the torque T_P generated by the pressure on inner helical structure surface with respect to the fixed end can be expressed as:

$$T_P = \int_H P_{in} \cdot b \cdot \sqrt{(2\pi R')^2 + z^2} \cdot z ds \tag{7}$$

where b is the width of the fiber cloth and H is the curve equation of the helix. The axial principal stress is the main internal stress for the bending of the actuator. For circular sections, the axial principal stretch ratio λ_2 is given by [30]:

$$\lambda_2 = \frac{L + \frac{r_2 - r_1}{\cos(\beta)(t_1+t_2)} \cdot \frac{bL}{d}}{L} = 1 + \frac{(r_2 - r_1) \cdot b}{\cos(\beta) \cdot (t_1 + t_2) \cdot d} \tag{8}$$

The tangential stress τ for twisting angle ϕ of a single actuator can be expressed by:

$$\tau = C_1 \lambda_2 \frac{r\phi}{L} + 2rC_2\lambda_2 \cos(\omega_0)(I-1)\left(\frac{\phi \cos(\omega_0)}{L} + \frac{\sin(\omega_0)}{R'}\right) \tag{9}$$

in which:

$$I = \lambda_2{}^2 \cos^2(\omega_0) + \frac{r_1{}^2 \phi^2 \cos^2(\omega_0)}{L^2} + 2\frac{r_1{}^2 \phi \cos(\omega_0) \sin(\omega_0)}{LR'} + \frac{r_1{}^2 \sin^2(\omega_0)}{R'^2} \tag{10}$$

where C_1 and C_2 are the material coefficients of the soft chamber and winding fiber, respectively [30]; ω_0 is the fiber cloth winding angle. Then, the torsional moment generated by the material can be calculated as:

$$T_i = \int\int_A \tau r_1 dA = 2\int_0^{\pi} d\alpha \int_{r_1}^{R} [C_1\lambda_2 \frac{r_1\phi}{L} + 2r_1 C_2 \lambda_2 \cos(\omega_0)(I_4 - 1)\left(\frac{\phi\cos(\omega_0)}{L} + \frac{\sin(\omega_0)}{R'}\right)] r_1 dl \tag{11}$$

Using $T_P = T_\phi$, the correlations between the torsional angle ϕ and the chamber pressure P_{in} can be solved for the single actuator. The twisting angle for the wrist can be considered equal to ϕ by neglecting the mutual resistance between the four chambers.

Summarizing, this section illustrated the approaches for estimating the relationships between input pressure and the bending and twisting angles. When the material and structure parameters of the soft wrist are determined, the bending and twisting angles can be calculated to assess the performance of the soft wrist under different applied air pressures.

4. Numerical Modeling

The previous section illustrated theoretical approaches for deriving bending and twisting angles of the soft wrist. To assess the accuracy of the theoretical analysis, finite element method simulations were carried out to predict the bending and twisting motions of a soft wrist whose size is comparable to a human wrist.

4.1. Simulation Parameters

The helical actuators are made of a soft silicone chamber and Kevlar fiber. The silicone rubbers are modeled as nonlinear elastic, isotropic, and incompressible under quasi-static loading. Finite element analyses (FEAs) are often conducted for modeling mechanical responses of the soft wrist. The simulations of the bending and twisting motion can be studied by modeling the soft wrist in the static structure module of ANSYS Workbench (2021R1) [31,32]. Using the hyperelastic incompressible Yeoh material model, the silicone chamber is modeled based on the theoretical Equation (12) [33,34]:

$$W = \sum_{i=1}^{3} c_{i0}\left(\overline{I} - 3\right)^{i} \tag{12}$$

where W is the strain energy; c_{i0} is a material constant (i = 1, 2, 3) and $\overline{I}$ is the first invariant. For the experimental materials of Ecoflex 20 [34], the material parameter values are available in the literature [31]. The simulation parameters and their values for modelling materials are listed in Table 2. In the simulations, the quality of the silicone cavity is defined by the hyperelastic incompressible Yeoh material model. The silicone chamber is meshed using Solid187, which is a 10-node tetrahedral element that is adaptable to the model to generate high-quality meshes [31]. The fiber is modeled with cable elements. Bonded constraints are applied for the contacts of the chambers and the contacts between chamber and fiber. The fixed constraint is applied to the bottom of the soft wrist.

Table 2. Simulation parameters for modeling materials.

Materials	Parameters	Values
Silicone chamber	Material constant C_1	0.1 MPa [30]
	Material constant C_2	0.013 MPa [30]
	Material constant C_3	0.00002 MPa [30]
Kevlar fiber	Young's modulus	28 GPa
	Poisson's ratio	0.37

4.2. Simulation Results

For simulating the behaviors of a soft wrist under pressurization, a reference case was selected based on the sizes of a human wrist. It was assumed that a human wrist has a circular cross section, and its radius is approximately 50 mm. The radius of each actuator R' can then be determined at 12 mm. Based on this radius, the actuator length L was set to be 58 mm because on the one hand, a shorter length will not produce sufficiently large bending or twisting angles for a test pressure of 50 kPa, and on the other hand, longer actuators for the wrist are not necessary. The determination of the chamber thickness and diameter also affected the fabrication conditions of the soft actuators. The finally determined sizes were: cylinder core radius r_1 = 5 mm, inner helix radius r_2 = 8 mm, inner helix width b = 2 mm, helix pitch d = 7 mm, outward layer thickness t_1 = 4 mm, and inward layer thickness t_2 = 2.25 mm. In the simulation, a pressure of 50 kPa is applied to the chambers of the four actuators. When adjacent chambers (i.e., *A*&*C*, *B*&*D*) are pressurized, side bending motions can be achieved, as given in Figure 5a. When diagonal chambers (i.e., *A*&*D*, *B*&*C*) are pressurized, the twisting motions are shown in Figure 5b. The results are consistent with the analysis of soft wrist motions in Table 1.

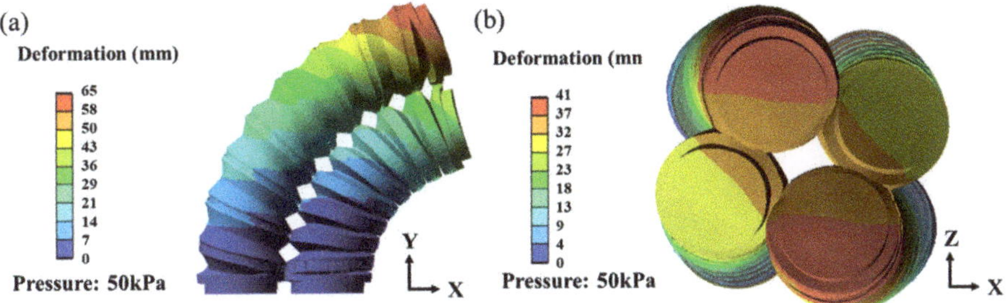

Figure 5. Soft wrist motion simulation: (**a**) bending right; (**b**) twisting right.

5. Experimental Tests

The previous sections have theoretically and numerically evaluated the bending and twisting motions of the soft wrist. To validate the predicted motions, a real sample of the soft wrist used in the simulation model was fabricated. The experimental tests of bending and twisting motions were compared with the theoretical and simulation results.

5.1. Sample Fabrication

The soft chambers were fabricated with Ecoflex-0020 liquid silicone (Smooth-on Inc., Macungie, PA, USA) using a casting process [35]. The molds used for the actuators were fabricated by a 3D printer (Objet Connex 500, Stratasys corporation, Eden Prairie, MN, USA).) using resin material. Equal amounts of part A and part B of Ecoflex-0020 were mixed and stirred until the mixture was evenly dispersed. Then, the mixture was placed in a vacuum chamber at room temperature for 3 min to remove bubbles. Subsequently, the mixture was poured into the molds and put in a vacuum drying oven for 3 h at 70 °C for thorough molding. Four Kevlar fibers of 0.5 mm diameter were manually wound along the helix structure of the actuator outer surface. After the four actuators with wound fibers were molded, the soft silicon pipes of 3 mm diameter were added to the chamber ends. Then, the same liquid silicone was poured onto the contacts between the four actuators, and the contacts between the actuators and the pipes. Finally, the actuators were placed in a vacuum drying oven for one hour at 70 °C. In this way, a model of a soft wrist can be fabricated.

5.2. Experimental Results

Figure 6 shows the fabricated sample and the experimental setup, which was composed of a micro air pump (370-B), an air pump pressure knob board, and a pressure gauge (HT1895). The micro air pump was used to supply air to the soft actuators. The air pump pressure knob can regulate the air flow of the air pump. The pressure gauge was used to measure the output pressure from the air pump [36]. Figure 7 illustrates the bending and twisting motions of the soft wrist under pressure of 50 kPa. The experimental results of increasing pressure on bending and twisting angles are plotted in Figure 8, in which the results from simulations and theoretical tests were also shown. The calculations for the theoretical results between input pressure and angles of bending and twisting are elaborated below.

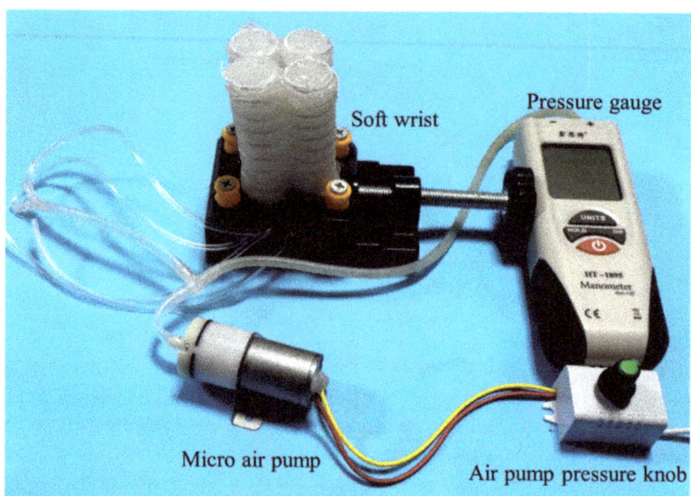

Figure 6. Experimental setup for measuring soft wrist motion.

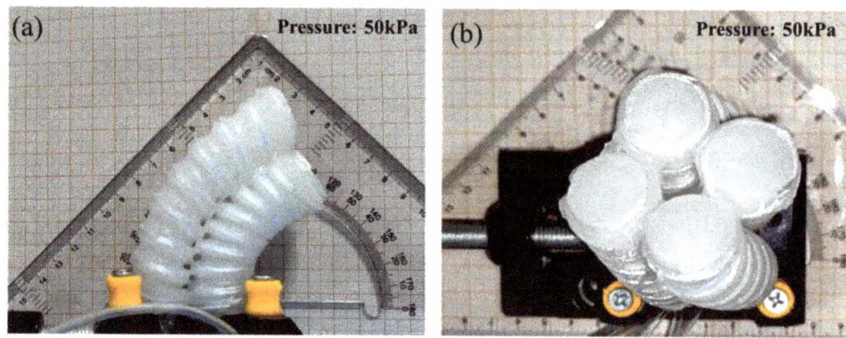

Figure 7. Soft wrist motion test: (**a**) bending; (**b**) twisting.

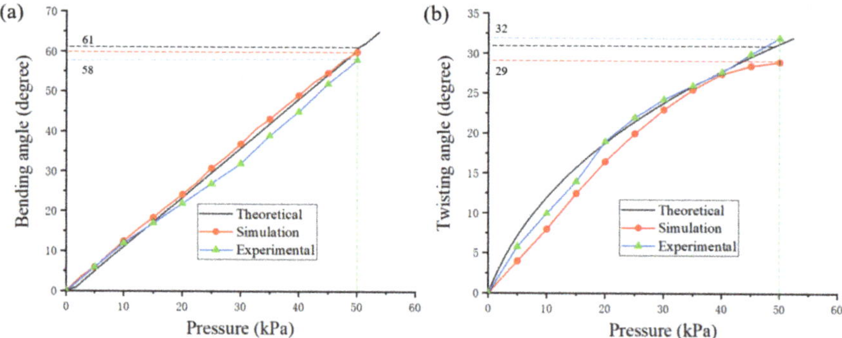

Figure 8. Theoretical, simulation, and experimental results for the soft wrist: (**a**) bending angle; (**b**) twisting angle.

For the fabricated sample shown in Figure 7, using the Equations (2) and (5) and the parameter values in Table 2, the relationship between the bending angle θ of the soft wrist and the pressure P_{in} can be estimated, which is given by below Equation (13). Likewise, using Equations (7) and (11), and the values in Table 2, the relationship between the twisting angle ϕ and the pressure P_{in} can be obtained, which is expressed by Equation (14):

$$P_{in} - 812\theta + 1000 + \frac{16}{\theta^3} = 0 \tag{13}$$

$$P_{in} - 0.5\phi^3 + 18\phi^2 + 550\phi = 0 \tag{14}$$

Figure 8 shows that overall, the bending and twisting angles of the soft wrist increased similarly with the increase of pressure for analytical, simulation, and experimental tests. The deviations between the three results are mainly ascribed to the simplified hyper-elastic model for the silicon chamber and the homogeneous assumptions for the restriction layers in the theoretical and simulation analysis [33]. To further illustrate that the deviations are small, the angles for bending and twisting for 50 kPa are denoted for the analytical, simulation, and experimental results. Both deviations of the bending and twisting angles are within 2.5%, which verifies that the simulation approach can be used to predict angles of bending and twisting for the soft wrist under pressure of 50 kPa.

6. Application for a Soft Hand

The previous section demonstrated that simulations are reliable for predicting angles of bending and twisting for the soft wrist under test pressures. In this section, the sensitivity of the geometric parameters on the bending and twisting of a soft wrist are studied using the simulation approach. Based on the optimal results, a series of values for the chamber structure were used to fabricate a soft wrist and implemented in the soft gripper.

6.1. Sensitivity Studies of Geometric Parameters

To obtain the maximal angles of bending and twisting while maintaining the length of the soft wrist, sensitivity studies of the chamber geometric parameters were conducted. To ensure that the chamber has sufficient space, the inner hole radius r_1 was not less than 4 mm. The inner helix radius r_2 was not greater than 10 mm referring to the size of a human wrist. The maximum helix width b was set to be 3 mm because these sizes are sufficient to restrain inflation of an actuator. Since the fiber is wound on the helix structure of the outer surface of each chamber, the winding angle is the same as the helix angle. Because the winding angle can be derived by the pitch d while the length L and radius r_2 remain the same, it is more straightforward to assess the pitch d than the winding angle to study its geometric influences on the angles of bending and twisting. The range for the helix pitch d was set as 5–9 mm, because smaller values result in difficulties in demolding. Taking the casting process into account, the outward layer thickness t_1 and inward layer thickness t_2 should be greater than 2 and 1.75 mm, respectively. The tested values for these geometric parameters for a chamber structure are listed in Table 3.

Table 3. Geometric parameters for chamber structure (unit: mm).

Inner Hole Radius r_1	Inner Helix Radius r_2	Helix Width b	Helix Pitch d	Outward Layer Thickness t_1	Inward Layer Thickness t_2
4	6	1	5	2	1.75
4.5	7	1.5	6	3	2
5	8	2	7	4	2.25
5.5	9	2.5	8	5	2.5
6	10	3	9	6	2.75

The influences of the geometric parameters on the angles of bending and twisting for air pressure of 50 kPa are presented in Figure 9. It can be seen from Figure 9a that both the bending and twisting angles are increased by increasing the inner hole radius r_1. This is because an increase in the inner radius reduces the thickness of the chamber, which leads to higher elasticity so that the deformation is larger [37]. Figure 9b demonstrates that larger helix radii r_2 result in smaller angles of bending and twisting. This is due to the fact that when the outward radius increases, the cross-sectional moment of inertia increases, leading to increased stiffness [38], thus resulting in a smaller actuator deformation. Figure 9c presents that increasing the width of the inner helix surface b enhances the motions of the soft wrist. The reason for this is that when the width of the inner helix surface increases, the force area inside the drive chamber increases, which also leads to greater deformation [39]. Figure 9d shows that the increase in pitch results in smaller bending and torsion angles. The reason can be ascribed to the fact that the increase in pitch leads to a reduction in the number of spiral turns of the actuator, which makes it less stretchable. Additionally, in the case of a constant drive length, since the number of spiral turns is inversely proportional to the pitch, the deformation capability of the actuator varies more when the pitch is small. Figure 9e,f illustrate that the increase of the inward thickness and the outward thickness of the chamber lead to an increase in the material of the actuator, resulting in an increase in its stiffness [38].

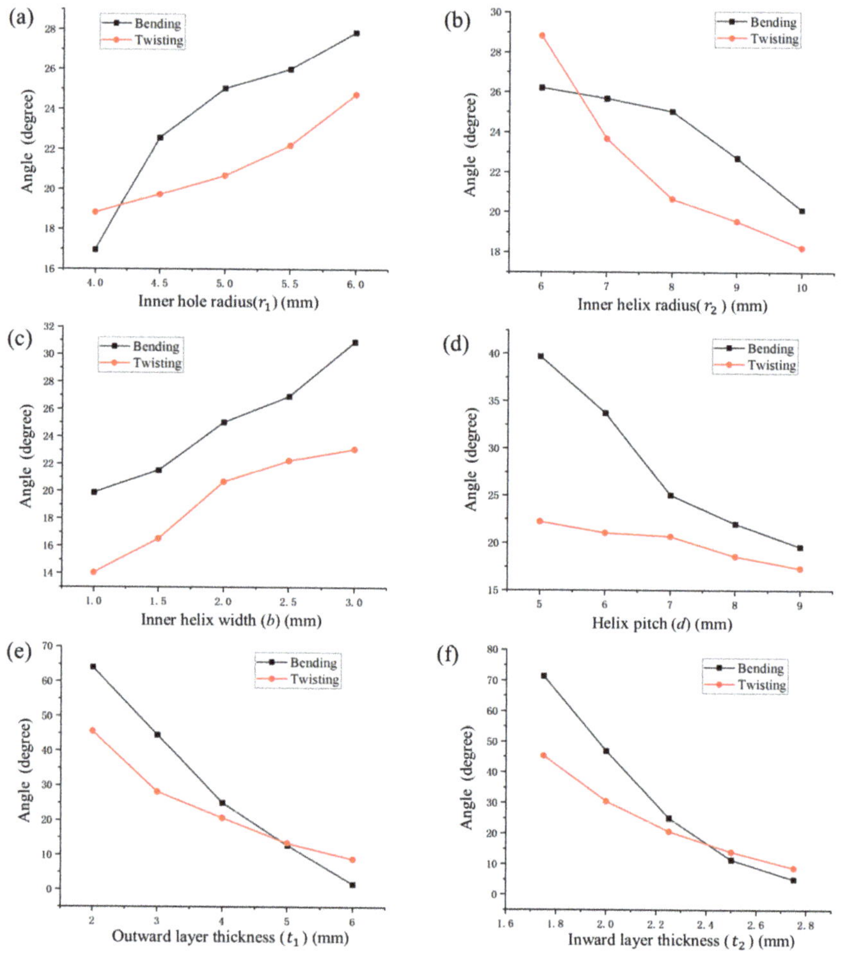

Figure 9. Analysis of structural parameters: (**a**) internal radius; (**b**) width of inner helix surface; (**c**) length; (**d**) pitch; (**e**) outer layer thickness; (**f**) inner layer thickness.

Summarizing, the sensitivity studies demonstrated that increasing the inner hole radius r_1, or the width of the inner helix surface b, will enable larger bending and twisting angles. By contrast, increasing the helix pitch d, outer radius r_2, outer layer thickness t_1, or inner layer thickness t_2 results in a decrease of both the bending and twisting angles. Besides the characterization of motion capability, the output forces, including blocking force and blocking torque, are two widely investigated parameters for characterizing the performances of a wrist. For measuring blocking force, the four actuators of the wrist will be installed on four holes of a rigid base support. The support will be fastened using bolts of a fixture as shown in Figure 6. In this way, one end of the wrist can be fixed and the other end can perform free motions upon pressurization. A force sensor with a conical tip [40] can be placed at the right side of the free end of the wrist. Upon pressurization and different geometric parameters, the corresponding blocking forces [41,42] and shearing forces [40] can be measured by the sensor. Knowing the length of the wrist, the blocking torque can be determined. Since this paper focuses on the evaluations of dexterous bending and twisting motions of an active wrist using pneumatic actuators, a thorough investigation of the geometric parameters' effects on the blocking forces and torques is planned in future work.

6.2. Verfication and Validation of the Soft Wrist

To obtain maximum angles of the bending and twisting, the cylinder core radius was selected at r_1 = 6 mm and the inner helix radius was set at r_2 = 8 mm based on the results of the sensitivity study. In combination with limitations in the casting process, other parameter values were determined, including: inner helix width b = 2 mm, pitch d = 6 mm, outer layer thickness $t_1 = 3$ mm, and inner layer thickness $t_2 = 2.25$ mm. Using the results of the simulations and the experimental tests, this set of values gives the bending angle and twisting angles of 54° and 23° for a pressure of 50 kPa. These results can well resemble the motions of a real human wrist, as shown in Figure 10. Therefore, this model of the soft wrist can improve grasping dexterity for the soft grippers. In combination with its lightweight driven system [9–11], promising applications of this wrist in soft robotics are forseen.

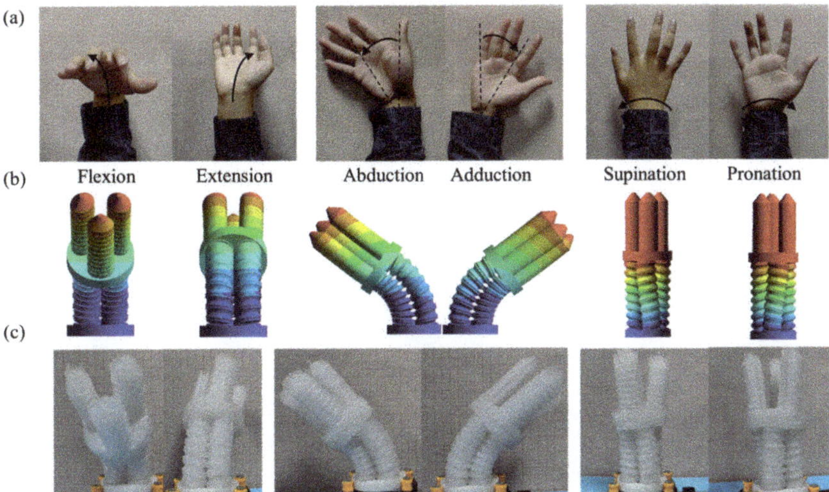

Figure 10. Motion of soft wrist for soft robotic hand: (**a**) real human hand; (**b**) simulation; (**c**) experiments.

7. Conclusions

This paper introduced an innovative structural model of a soft wrist used for soft grippers. It can achieve both bending and twisting motions for one segment, which is a unique feature. This soft wrist is also lightweight and energy-saving because one simple air pump is used in the driving system. Kinetic analysis, simulations, and experimental tests have all demonstrated that the soft wrist can produce the desired bending and twisting motions. The sensitivity studies demonstrated that increasing the inner hole radius r_1, or the width of the inner helix surface b, can enable larger bending and twisting angles. On the contrary, increasing helix pitch d, outer radius r_2, outer layer thickness t_1, or inner layer thickness t_2 results in a decrease of both the bending and twisting angles. The implementation of a fabricated soft wrist on a soft robotic hand showed that the soft wrist can improve grasping dexterity for soft grippers.

Author Contributions: G.C. and A.J. conceptualized and reviewed the manuscript; The experimental methodology and sample fabrications were accomplished by T.L.; G.L. reviewed and edited the manuscript; the data processing and the original draft preparation were accomplished by S.D. and S.C. All authors have read and agreed to the published version of the manuscript.

Funding: This research was funded by the Foundation Research Project of Jiangsu Province Natural Science Fund (No. BK20190415).

Data Availability Statement: The datasets generated during and/or analyzed during the current study are available from the corresponding author upon reasonable request.

Conflicts of Interest: The authors declare no conflict of interest.

Nomenclature

A	Force integral area (m^2)
b	Width of inner helix surface (m)
c	Length of the chamber internal structure (m)
d	Pitch (m)
c_{i0}	Material constant (Pa)
C_1	Silicone rubber material parameter (Pa)
C_2	Fiber material parameter (Pa)
D	Actuator diameter (m)
D_k	Incompressibility parameter (Pa^{-1})
e	Thickness of fiber layer (m)
f	Force generated by air pressure (N)
G_0	Initial shear modulus (Pa)
h	Distance from bottom edge (m)
I	Parameter in tangential stress calculation (None)
$\bar{I}$	The first invariant (None)
J	Volume variation ratio (None)
L	Actuator length (m)
M_P	Moment generated by pressure (bending) (N·m)
M_i	Moment generated by material (bending) (N·m)
M_θ	Moment generated by the material (N·m)
P_{in}	Chamber pressure (pa)
r	Actuator chamber radius (m)
r_1	Internal radius (m)
r_2	Outer radius (m)
R	Curvature radius
R'	Actuator radius (m)
t	Average equivalent thickness (m)
t_1	Outer layer thickness (m)
t_2	Inner layer thickness (m)
T_P	Moment generated by pressure (twisting) (N·m)

T_i	Moment generated by material (twisting) (N·m)
β	Helix angle (rad)
θ	Bending angle (rad)
λ_1	Stretch ratio in bending model (none)
λ_2	Stretch ratio in twisting model (none)
τ	Shear stress (Pa)
ω	Angle of fiber cloth (rad)
ω_0	Helix angle of fiber cloth (rad)
ϕ	Twist angle (rad)

References

1. Li, M.; Pal, A.; Aghakhani, A.; Pena-Francesch, A.; Sitti, M. Soft Actuators for Real-World Applications. In *Nature Reviews Materials;* Springer: New York, NY, USA, 2022; pp. 235–249.
2. Wang, D.; Wu, X.; Zhang, J.; Du, Y. A Pneumatic Novel Combined Soft Robotic Gripper with High Load Capacity and Large Grasping Range. *Actuators* **2022**, *11*, 3. [CrossRef]
3. Grazioso, S.; Di Gironimo, G.; Siciliano, B. A Geometrically Exact Model for Soft Continuum Robots: The Finite Element Deformation Space Formulation. *Soft Robot.* **2019**, *6*, 790–811. [CrossRef] [PubMed]
4. Lu, M.; Chen, G.; He, Q.; Zong, W.; Yu, Z.; Dai, Z. Development of a Hydraulic Driven Bionic Soft Gecko Toe. *J. Mech. Robot.* **2021**, *13*, 051005. [CrossRef]
5. Zhang, Z.; Ni, X.; Wu, H.; Sun, M.; Bao, G.; Wu, H.; Jiang, S. Pneumatically Actuated Soft Gripper with Bistable Structres. *Soft Robot.* **2022**, *9*, 57–71. [CrossRef] [PubMed]
6. Zhang, Z.; Ni, X.; Gao, W.; Shen, H.; Sun, M.; Guo, G.; Wu, H.; Jiang, S. Pneumatically Controlled Reconfigurable Bistable Bionic Flower for Robotic Gripper. *Soft Robot.* **2022**, *9*, 657–669. [CrossRef] [PubMed]
7. Gloumakov, Y.; Spiers, A.J.; Dollar, A.M. Dimensionality Reduction and Motion Clustering during Activities of Daily Living: Decoupling Hand Location and Orientation. *IEEE Trans. Neural Syst. Rehabil. Eng.* **2020**, *28*, 2955–2965. [CrossRef]
8. El-Atab, N.; Mishra, R.B.; Al-Modaf, F.; Joharji, L.; Alsharif, A.A.; Alamoudi, H.; Diaz, M.; Qaiser, N.; Hussain, M.M. Soft Actuators for Soft Robotic Applications: A Review. *Adv. Intell. Syst.* **2020**, *2*, 2000128. [CrossRef]
9. Abondance, S.; Teeple, C.B.; Wood, R.J. A Dexterous Soft Robotic Hand for Delicate In-Hand Manipulation. *IEEE Robot. Autom. Lett.* **2020**, *5*, 5502–5509. [CrossRef]
10. Subramaniam, V.; Jain, S.; Agarwal, J.; Valdivia y Alvarado, P. Design and Characterization of a Hybrid Soft Gripper with Active Palm Pose Control. *Int. J. Rob. Res.* **2020**, *39*, 1668–1685. [CrossRef]
11. Zhong, G.; Hou, Y.; Dou, W. A Soft Pneumatic Dexterous Gripper with Convertible Grasping Modes. *Int. J. Mech. Sci.* **2019**, *153–154*, 445–456.
12. Zhang, H.; Kumar, A.S.; Fuh, J.Y.H.; Wang, M.Y. Design and Development of a Topology-Optimized Three-Dimensional Printed Soft Gripper. *Soft Robot.* **2018**, *5*, 650–661. [CrossRef]
13. Gong, Z.; Fang, X.; Chen, X.; Cheng, J.; Xie, Z.; Liu, J.; Chen, B.; Yang, H.; Kong, S.; Hao, Y.; et al. A Soft Manipulator for Efficient Delicate Grasping in Shallow Water: Modeling, Control, and Real-World Experiments. *Int. J. Rob. Res.* **2021**, *40*, 449–469. [CrossRef]
14. Xiao, W.; Hu, D.; Chen, W.; Yang, G.; Han, X. Modeling and Analysis of Bending Pneumatic Artificial Muscle with Multi-Degree of Freedom. *Smart Mater. Struct.* **2021**, *30*, 095018. [CrossRef]
15. Alici, G.; Canty, T.; Mutlu, R.; Hu, W.; Sencadas, V. Modeling and Experimental Evaluation of Bending Behavior of Soft Pneumatic Actuators Made of Discrete Actuation Chambers. *Soft Robot.* **2018**, *5*, 24–35. [CrossRef]
16. Sanan, S.; Lynn, P.S.; Griffith, S.T. Pneumatic Torsional Actuators for Inflatable Robots. *J. Mech. Robot.* **2014**, *6*, 031003. [CrossRef]
17. Chatterjee, A.; Chahare, N.R.; Kondaiah, P.; Gundiah, N. Role of Fiber Orientations in the Mechanics of Bioinspired Fiber-Reinforced Elastomers. *Soft Robot.* **2021**, *8*, 640–650. [CrossRef]
18. Chandler, J.H.; Chauhan, M.; Garbin, N.; Obstein, K.L.; Valdastri, P. Parallel Helix Actuators for Soft Robotic Applications. *Front. Robot. AI* **2020**, *7*, 119. [CrossRef]
19. Jiang, C.; Wang, D.; Zhao, B.; Liao, Z.; Gu, G. Modeling and Inverse Design of Bio-Inspired Multi-Segment Pneu-Net Soft Manipulators for 3D Trajectory Motion. *Appl. Phys. Rev.* **2021**, *8*, 041416. [CrossRef]
20. Nassour, J. Marionette-Based Programming of a Soft Textile Inflatable Actuator. *Sensors Actuators, A Phys.* **2019**, *291*, 93–98. [CrossRef]
21. Xavier, M.S.; Tawk, C.D.; Zolfagharian, A.; Pinskier, J.; Howard, D.; Young, T.; Lai, J.; Harrison, S.M.; Yong, Y.K.; Bodaghi, M.; et al. Soft Pneumatic Actuators: A Review of Design, Fabrication, Modeling, Sensing, Control and Applications. *IEEE Access* **2022**, *10*, 59442–59485. [CrossRef]
22. Kurumaya, S.; Phillips, B.T.; Becker, K.P.; Rosen, M.H.; Gruber, D.F.; Galloway, K.C.; Suzumori, K.; Wood, R.J. A Modular Soft Robotic Wrist for Underwater Manipulation. *Soft Robot.* **2018**, *5*, 399–409. [CrossRef]
23. Fei, Y.; Wang, J.; Pang, W. A Novel Fabric-Based Versatile and Stiffness-Tunable Soft Gripper Integrating Soft Pneumatic Fingers and Wrist. *Soft Robot.* **2019**, *6*, 1–20. [CrossRef] [PubMed]

24. Shen, Z.; Zhong, H.; Xu, E.; Zhang, R.; Yip, K.C.; Chan, L.L.; Chan, L.L.; Pan, J.; Wang, W.; Wang, Z. An Underwater Robotic Manipulator with Soft Bladders and Compact Depth-Independent Actuation. *Soft Robot.* **2020**, *7*, 535–549. [CrossRef] [PubMed]
25. Wang, T.; Ge, L.; Gu, G. Programmable Design of Soft Pneu-Net Actuators with Oblique Chambers Can Generate Coupled Bending and Twisting Motions. *Sens. Actuators A Phys.* **2018**, *271*, 131–138. [CrossRef]
26. Haddad, F.S. Automatic Design of Fiber-Reinforced Soft Actuators. *Proc. Natl. Acad. Sci. USA* **2017**, *114*, 51–56.
27. Sui, D.; Zhao, S.; Wang, T.; Liu, Y.; Zhu, Y.; Zhao, J. Design of a Bio-Inspired Extensible Continuum Manipulator with Variable Stiffness. *J. Bionic Eng.* **2022**. [CrossRef]
28. Tian, Y.; Zhang, Q.; Cai, D.; Chen, C.; Zhang, J.; Duan, W. Theoretical Modelling of Soft Robotic Gripper with Bioinspired Fibrillar Adhesives. *Mech. Adv. Mater. Struct.* **2020**, *29*, 2250–2266. [CrossRef]
29. Wang, Z.; Polygerinos, P.; Overvelde, J.T.B.; Galloway, K.C.; Bertoldi, K.; Walsh, C.J. Interaction Forces of Soft Fiber Reinforced Bending Actuators. *IEEE/ASME Trans. Mechatronics* **2017**, *22*, 717–727. [CrossRef]
30. Sedal, A.; Bruder, D.; Bishop-Moser, J.; Vasudevan, R.; Kota, S. A Continuum Model for Fiber-Reinforced Soft Robot Actuators. *J. Mech. Robot.* **2018**, *10*, 024501. [CrossRef]
31. Xavier, M.S.; Fleming, A.J.; Yong, Y.K. Finite Element Modeling of Soft Fluidic Actuators: Overview and Recent Developments. *Adv. Intell. Syst.* **2021**, *3*, 2000187. [CrossRef]
32. ANSYS®Inc. *ANSYS®Mechanical User's Guide*; ANSYS®: Canonsburg, PA, USA, 2021.
33. Chen, L.; Yang, C.; Wang, H.; Branson, D.T.; Dai, J.S.; Kang, R. Design and Modeling of a Soft Robotic Surface with Hyperelastic Material. *Mech. Mach. Theory* **2018**, *130*, 109–122. [CrossRef]
34. Marechal, L.; Balland, P.; Lindenroth, L.; Petrou, F.; Kontovounisios, C.; Bello, F. Toward a Common Framework and Database of Materials for Soft Robotics. *Soft Robot.* **2021**, *8*, 284–297. [CrossRef]
35. Rus, D.; Tolley, M.T. Design, Fabrication and Control of Soft Robots. *Nature* **2015**, *521*, 467–475. [CrossRef]
36. Cacucciolo, V.; Renda, F.; Poccia, E.; Laschi, C.; Cianchetti, M. Modelling the Nonlinear Response of Fibre-Reinforced Bending Fluidic Actuators. *Smart Mater. Struct.* **2016**, *25*, 105020. [CrossRef]
37. Guo, D.; Kang, Z. Chamber Layout Design Optimization of Soft Pneumatic Robots. *Smart Mater. Struct.* **2020**, *29*, 025017. [CrossRef]
38. Webster, R.J.; Jones, B.A. Design and Kinematic Modeling of Constant Curvature Continuum Robots: A Review. *Int. J. Rob. Res.* **2010**, *29*, 1661–1683. [CrossRef]
39. Xiao, W.; Hu, D.; Yang, G.; Jiang, C. Modeling and Analysis of Soft Robotic Surfaces Actuated by Pneumatic Network Bending Actuators. *Smart Mater. Struct.* **2022**, *31*, 055001. [CrossRef]
40. Chen, F.; Miao, Y.; Gu, G.; Zhu, X. Soft Twisting Pneumatic Actuators Enabled by Freeform Surface Design. *IEEE Robot. Autom. Lett.* **2021**, *6*, 5253–5260. [CrossRef]
41. Polygerinos, P.; Wang, Z.; Overvelde, J.T.B.; Galloway, K.C.; Wood, R.J.; Bertoldi, K.; Walsh, C.J. Modeling of Soft Fiber-Reinforced Bending Actuators. *IEEE Trans. Robot.* **2015**, *31*, 778–789. [CrossRef]
42. Zhang, X.; Oseyemi, A.E. A Herringbone Soft Pneu-Net Actuator for Enhanced Conformal Gripping. *Robotica* **2022**, *40*, 1345–1360. [CrossRef]

 *actuators* 

Article

Effects of Flexural Rigidity on Soft Actuators via Adhering to Large Cylinders

Liuwei Wang [1,2,†], Qijun Jiang [1,†], Zhiyuan Weng [1,2], Qingsong Yuan [1] and Zhouyi Wang [1,2,*]

1 College of Mechanical and Electrical Engineering, Nanjing University of Aeronautics and Astronautics, Nanjing 210016, China
2 Nanjing University of Aeronautics and Astronautics Shenzhen Research Institute, Shenzhen 518063, China
* Correspondence: wzyxml@nuaa.edu.cn
† These authors contributed equally to this work.

Abstract: This study proposes a soft pneumatic actuator with adhesion (SPAA) consisting of a top fluidic-driven elastic actuator and four bottom adhesive pads for adhering to large cylinders. Finite element models were developed to investigate the bending properties under positive air pressure and the effect of "rib" height on the flexural rigidity of the SPAA. A synchronous testing platform for the adhesive contact state and mechanics was developed, and the bending curvature and flexural rigidity of the SPAA were experimentally measured relative to the pressure and "rib" height, respectively, including the adhesion performance of the SPAA with different rigidities on large cylinders. The obtained results indicate that the SPAA can continuously bend with controllable curvature under positive air pressure and can actively envelop a wide range of cylinders of different curvatures. The increase in the "rib" height from 4 to 8 mm increases the flexural rigidity of the SPAA by approximately 230%, contributing to an average increase of 54% in the adhesion performance of the SPAA adhering to large cylinders. The adhesion performance increases more significantly with an increase in the flexural rigidity at a smaller peeling angle. SPAA has a better adhesion performance on large cylinders than most existing soft adhesive actuators, implying that is more stable and less affected by the curvature of cylinders. To address the low contact ratio of the SPAA during adhesion, the optimization designs of the rigid–flexible coupling hierarchical and differentiated AP structures were proposed to increase the contact ratio to more than 80% in the simulation. In conclusion, this study improved the adhesion performance of soft adhesive actuators on large cylinders and extended the application scope of adhesion technology. SPAA is a basic adhesive unit with a universal structure and large aspect ratio similar to that of the human finger. According to working conditions requirements, SPAAs can be assembled to a multi-finger flexible adhesive gripper with excellent maneuverability.

Keywords: flexural rigidity; soft actuator; adhesion; large cylinders

Citation: Wang, L.; Jiang, Q.; Weng, Z.; Yuan, Q.; Wang, Z. Effects of Flexural Rigidity on Soft Actuators via Adhering to Large Cylinders. *Actuators* **2022**, *11*, 286. https://doi.org/10.3390/act11100286

Academic Editors: Hamed Rahimi Nohooji and Steve Davis

Received: 12 September 2022
Accepted: 1 October 2022
Published: 7 October 2022

1. Introduction

Gripping large smooth cylinders is challenging for robotic grippers [1,2]. Conventional grippers use normal and frictional forces to grip objects. If an object is relatively smaller than the gripper, the gripper can envelop the object to form a shape closure that holds it; however, when the object is relatively larger than the gripper, the gripper squeezes the surface of the object to generate frictional forces and grips to lift the object while relying on the resultant force of normal squeezing and tangential frictional forces. In this case, the component force of each squeezing force is vertically downward, and this tends to push the object away from the gripper. As the object becomes larger, the gripper needs to open wider to fit it, and this increases the resultant downward force (sum of the downward component force of each squeezing force). When the downward resultant force exceeds the upward lifting force, the gripper is unable to grip the object [3]. Therefore, interfacial adhesion technologies, such as vacuum adsorption, electrostatic adhesion, and gecko-like dry adhesion, which can generate interfacial attraction without excessive squeezing, have been employed in grippers [2]. Vacuum adsorption can provide a significant lifting force on non-porous objects; however, it is unsuitable for low-pressure and underwater environments [4]. Electrostatic adhesion can produce active adhesion on smooth and rough surfaces; however, it has a limited performance on non-metallic objects and requires additional control infrastructure to provide kV level high voltages [5]. Gecko-like dry adhesions rely on the van der Waals force generated by the close contact between the gecko-inspired micro-structured surface and the target to maintain adhesion. Moreover, various adhesive grippers have been developed for different applications.

According to the working conditions requirements, the adhesive gripper is assembled using several adhesive actuators; therefore, the mechanical properties of an adhesive actuator directly determine the operating performance of the gripper. For instance, rigid grippers assembled using two or three rigid planar adhesive actuators [6–9] exhibit excellent gripping performance on flat objects (PCB boards, glass sheets, and silicon wafers) but are not suitable for curved objects (cylinders); while soft grippers assembled using tendon-driven under-actuated adhesive actuators [5,10–12] or fluid-driven elastic actuators [13,14] exhibit excellent performance in gripping cylinders. However, the lifting capacity of soft adhesive grippers when adhering to cylinders decreases obviously with an increase in the radius of the cylinder. For adhering to large cylinders, the structural rigidity evidently affects the performance of the adhesive actuator/unit [4,15]. Although the inherent compliance of soft adhesive actuators can increase their adaptability to a wide range of targets of different sizes, they also reduce their structure rigidity, which is essential for maintaining adhesion, thereby reducing the peak adhesion performance [16,17]. By embedding photoresistant materials with a high elastic module into an elastic adhesive unit, the adhesion force can be enhanced [18]. In addition, the rigid–flexible adhesive unit can generate a larger area of adhesion during loading than a single flexible adhesive unit, resulting in a higher normal adhesion force and the ratio of adhesion force to preload [19]. As to the fluidic–elastic actuator, it has different flexural rigidity at different positions along its axial direction [13]. Moreover, the effective length of the fluidic–elastic actuator significantly affects the lifting force of the gripper [20], indicating that the structural stiffness of the actuator can be optimized to improve the performance of the gripper on objects with different curvatures. In general, the dependence of the adhesion on the structural parameters of the soft actuator is a crucial issue. The rigidity of the soft adhesive actuator/unit not only determines the acting mechanism of the actuator on the underlying adhesion system, but also affects the adhesion properties of the actuator adhering to the targets. Therefore, to adhere to large cylinders, it is essential to explore the effect of structural rigidity on the adhesion property of the soft actuator, which is helpful in guiding the mechanical design and rigidity optimization of the soft adhesive actuator to improve the adhesion performance on large cylinders.

This study proposes a soft pneumatic actuator with adhesion (SPAA) that combines a fluidic-driven elastic actuator with a mushroom-shaped microstructured adhesive surface for adhering to large cylinders. The effect of the flexural rigidity of the SPAA when it adheres to large cylinders is investigated. Section 2 presents the design of SPAAs with different bending rigidities and establishes finite-element models (FEMs) to predict their bending and rigidity properties. Section 3 describes the fabrication of the SPAAs and establishment of the experimental platform. Section 4 experimentally evaluates the performance of SPAAs with different bending rigidities when adhering to large cylinders. Section 5 compares SPAAs with existing adhesive actuators and proposes structural optimization schemes to solve the challenge of insufficient contact through finite element analysis (FEA). Section 6 provides a brief conclusion and potential areas for future studies.

2. Design and FEMs of the SPAA

2.1. Design of the SPAA

The SPAA (Figure 1a) consists of a top fluidic-driven elastic actuator and four bottom adhesive pads (Figure 1b). Specifically, the elastic actuator comprises four pneumatic networks [21,22] (PN, numbered PN1~PN4 from the root to end) connected through the inside air channel and an inextensible but flexible layer. As shown in Figure 1c, when a pressure P applies to the inner side wall of the PN, the resultant force acts on the geometric center of the side wall. Due to the offset between the resultant force and the inextensible layer, each PN is subjected to a resultant moment, causing the SPAA to bend to the bottom. Each PN has a length of 30 mm, a width of 9 mm, a height of 18 mm, and a thickness of 3 mm. Four adhesive pads (AP, numbered AP1~AP4 from the root to the end) are equidistantly bonded on the bottom of the SPAA. Each AP consists of the foamed rubber and a mushroom-shaped microstructured dry adhesive surface [23] with a length of 30 mm, a width of 10 mm, and a thickness of 3 mm. The SPAA has a total length of 48 mm under no deformation.

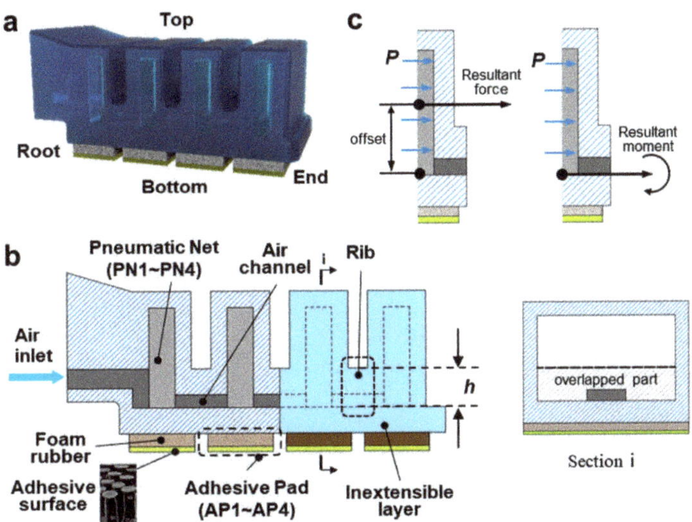

Figure 1. (**a**) Design of the SPAA. (**b**) Schematic illustration. (**c**) Actuation mechanism.

In order to study the effect of the flexural rigidity (EI) of the SPAA when it adheres to large cylinders, we aimed to obtain a series of SPAAs with significant differences in EI by changing the geometric parameters of the "rib" (the connection part between two PNs (Figure 1b)). The reference [21] proposed that the larger the rib width, the lower the EI, but the effect is not apparent. In this study, the SPAA would generate a slight forced-reverse bending when adhering to a cylinder, while the pressure on the inner sidewall of PNs resists this forced bending. In particular, the resistance directly acts on the overlapped part of the inner sidewall (section I in Figure 1b) and the rib, so we assumed that the area of this overlapped part has a significant effect on the EI of the SPAA. Finally, we designed three SPAAs with the rib height h of 4 mm, 6 mm, and 8 mm, and the rest of the geometric parameters were the same.

2.2. FEMs of the Bending Curvature and Flexural Rigidity

To predict the bending curvature and flexural rigidity of the SPAA under positive pressure, FEMs were established and analyzed using Abaqus/Standard (SIMULIA, Dassault System, Providence, RI, USA). The hyper-elastic Mooney–Rivlin model [20,24] (C_{10} = 0.4138, C_{01} = 0.1034, d = 2) and the Arruda–Boyce model [25] (μ = 0.5788, λ_m = 1.2099) were used to characterize the top elastic actuator and the bottom foam rubber, respectively.

Figure 2a shows the FEM of the bending curvature of the SPAA. The end of the SPAA was fixed, and a positive air pressure P (varying from 0 to 100 kPa at 10 kPa intervals) was uniformly applied to the inner wall. In order to accurately describe the bending state of the SPAA, we used C_1~C_4 to characterize the bending curvature of PN1~PN4, respectively.

We divided the abdomen of the SPAA into four equal parts with five yellow markers (numbered marker0~marker4 from the root to the end), recorded the simulation results of the five markers' coordinates under the SPAA's bending state, and obtained C_1~C_4 according to $C = \alpha/l$ Where α and l, respectively, represent the central angle and length of the arc that each PN bends into under positive pressure. Specifically, α can be obtained by geometric calculation based on the recorded data of markers' coordinates (the angles between the connecting line of Marker0 and Marker1~Marker4 and the vertical direction were numbered as θ_1~θ_4, respectively, then α_1~α_4 were $2\theta_1$, $2(\theta_2 - \theta_1)$, $2(\theta_3 - \theta_2)$, $2(\theta_4 - \theta_3)$, respectively). And l = 12 mm. The simulation results of four PNs' bending curvatures under the pressures varying from 0 to 100 kPa at 10 kPa intervals (Figure 2b) show that C_1~C_4 all increase with an increase in the air pressure, and the relationships are approximately linear ($R^2_{C_1} = 0.9974$, $R^2_{C_2} = 0.9978$, $R^2_{C_3} = 0.9963$, $R^2_{C_4} = 0.9970$). However, C_1 is obviously lower than C_2~C_4. It is because one side of PN1 is directly fixed to the root, thus limiting the bending of the PN1. As for PN2~PN4, the bending curvature is similar. Therefore, we used the mean value of the C_2~C_4 to define the bending curvature C of the SPAA. The simulation results of bending curvatures of three SPAAs with different rib heights under pressures varying from 0 to 100 kPa at 10 kPa intervals (Figure 2c) show that bending curvatures of SPAAs are positively correlated with the pressure, and the relationships are approximately linear ($R^2_{h=4} = 0.9957$, $R^2_{h=6} = 0.9980$, $R^2_{h=8} = 0.9996$). The results indicate that the SPAA can continuously bend with controllable bending curvature under positive air pressures.

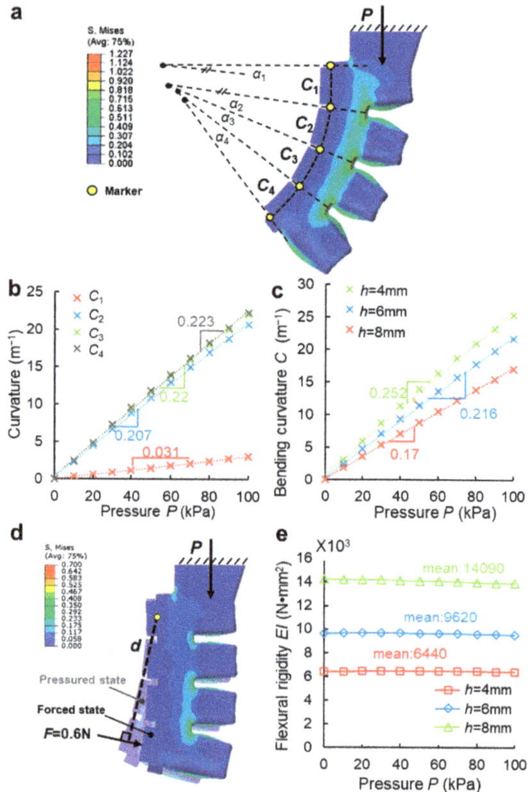

Figure 2. (**a**) The finite element model of the bending curvature of the SPAA. (**b**) The curvature of four PNs when the SPAA under pressure drive. (**c**) The bending curvature of three SPAAs with different rib heights under pressure drive. (**d**) The finite element model of the flexural rigidity of the SPAA. (**e**) The flexural rigidity of three SPAAs with different rib heights under pressure drive.

Figure 2d shows the FEM of the flexural rigidity of the SPAA. Based on the bending curvature model in Figure 2a, we applied a normal force $F = 0.6$ N to the end of the deformed SPAA, recorded the simulation results of the five markers' coordinates, and calculated the change of curvature ΔC of the SPAA before and after the action. According to reference [21], we defined the flexural rigidity $EI = Fd/\Delta C$ to represent the ability of the SPAA to resist the bending deformation, where d represents the distance from the root of SPAA to the force F. The simulation results of flexural rigidity of three SPAAs with different rib heights under pressures varying from 0 to 100 kPa at 10 kPa intervals (Figure 2e) show that the flexural rigidity is almost unaffected by the pressure, but is significantly affected by the height of rib. When the height of the rib increases from 4 to 8 mm, the mean value of the flexural rigidity increases from 6440 to 14,090 N·mm^2, an increase of 120%, which is much higher than the 10% increase in the flexural rigidity generated by decreasing the width of the rib from 3 to 0.8 mm [21]. The results indicate that it is theoretically reasonable to significantly increase the flexural rigidity of the SPAA by increasing the height of the rib.

3. Experimental Setup for Characterizing the Performance of the SPAA

3.1. Fabrication of the SPAA

The top elastic actuator was integrally formed by flexible 3D printing using SLA technology, which is simpler and more efficient than conditional pouring processing methods. The machine used is Form3 (Formlabs, Boston, MA, USA). The material used is flexible resin elastic 50 A with an elastic modulus of 2.89 MPa and a hardness of 50A after curing [26]. The foam rubbers were equidistantly attached to the bottom of the actuator using a 3M tape, and the dry adhesive surfaces [23] were attached to the foam rubbers. Three SPAA samples with rib heights of 4, 6, and 8 mm are shown in Figure 3d.

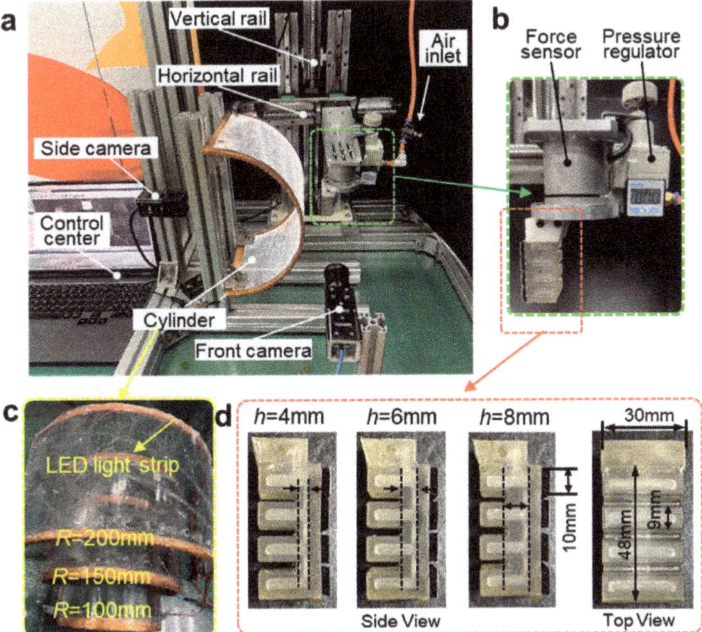

Figure 3. Experimental setup. (**a**) The synchronous testing platform. (**b**) Assembly layout of the SPAA, the six-dimensional force sensor and the pressure regulator. (**c**) Three acrylic semi-cylinders with radii of 100 mm, 150 mm and 200 mm. (**d**) Three SPAA samples prepared by 3D printing and their structural parameters.

3.2. Setup of the Synchronous Testing Platform

The synchronous testing platform of adhesive contact state and mechanics, as shown in Figure 3a, was built to evaluate the deformation, rigidity, and adhesion-peeling performance of the SPAA on large cylinders. The SPAA was fixed on a two-axis mobile platform that can move at any angle within 0~100 mm/s in the plane through a rigid cantilever beam. A six-dimensional force sensor (NBIT, Nanjing, China) was installed between the SPAA and the cantilever beam to record the force value during the SPAA adhering to cylinders. The force sensor has a force range of ±100 N, a moment range of ±5 N·m, and a resolution of 0.10% F·S. A manual pressure regulator (ZKAY, WuXi, China) that can continuously adjust the pressure in the range of 0~0.2 MPa was series connected in the gas circuit to adjust the driving pressure. In order to avoid the pressure pulsation caused by the sudden change of pressure, a lower pressure regulation speed was adopted. The digital indicator showed the internal pressure value of the SPAA in real-time with a resolution of 1 kPa (Figure 3b). The smooth and light-transmitting acrylic cylinder with LED light strips illuminating the edges (Figure 3c). According to the principle of frustrated total reflection [27], the area where the SPAA contacts the cylinder generated a facula with a significantly higher brightness than the non-contact area, which is convenient for extracting the contact area by computer graphics processing. The large cylinders with a radius of approximately two (R = 100 mm), three (R = 150 mm), and four times (R = 200 mm) of the length of the SPAA were selected for SPAAs to adhere to. Markers were drawn with a highlighter on one side of the SPAA in order to extract the coordinate information of Markers through computer graphics processing to calculate the bending curvature of the SPAA. Two high-speed cameras were placed on the front and side of the platform, respectively. The front recorded the bending and contact states of the SPAA, and the side recorded the distribution of the contact area.

4. Results

4.1. Bending Curvature and Flexural Rigidity

Figure 4a shows the bending state of three SPAAs with rib heights of 4, 6, and 8 mm at 0, 50, and 100 kPa, respectively. With the increase in pressure, three SPAAs all bend toward the bottom, and the curvature increases. Figure 4b shows the experimental results of the bending curvature of three SPAAs at pressures varying from 0 to 100 kPa at 10 kPa intervals. It is obvious that the bending curvatures of three SPAAs are positively correlated with the pressure, and the relationships are approximately linear ($R^2_{h=4} = 0.9953$, $R^2_{h=6} = 0.9982$, $R^2_{h=8} = 0.9995$). When P = 100 kPa, with an increase in the rib height, the bending curvatures of three SPAAs reach 25.4, 21.2, and 16.5 m^{-1}, respectively. It can be verified that the SPAA can continuously bend under positive air pressure drive and the bending curvature is controllable. Compared to the rigid [9,28] or under-actuated [10] soft actuators, SPAAs can actively envelop a wide range of cylinders with different curvatures. The simulation results in Figure 2c deviate within 7% of the experimental results.

Figure 4c shows the test flow of SPAA's flexural rigidity, which is similar to the FEM in Figure 2d. Figure 4d shows the experimental results of the flexural rigidity of three SPAAs under pressures varying from 0 to 100 kPa at 10 kPa intervals. The rigidity increases when the rib height increases but remains almost constant when the pressure increases. As the rib height goes up from 4 to 8 mm, the flexural rigidity of the SPAA increases from 6125 to 13,300 N·mm^2, an increase of 117%, demonstrating that it is more reasonable to significantly increase the flexural rigidity of the SPAA by increasing the rib height than decreasing the rib width [21]. The above experimental and FEM results in Figure 2 have the same trend with a high degree of agreement, and the deviations are all within 10%. Therefore, the FEMs in Section 2.2 can predict the mechanical properties of SPAA.

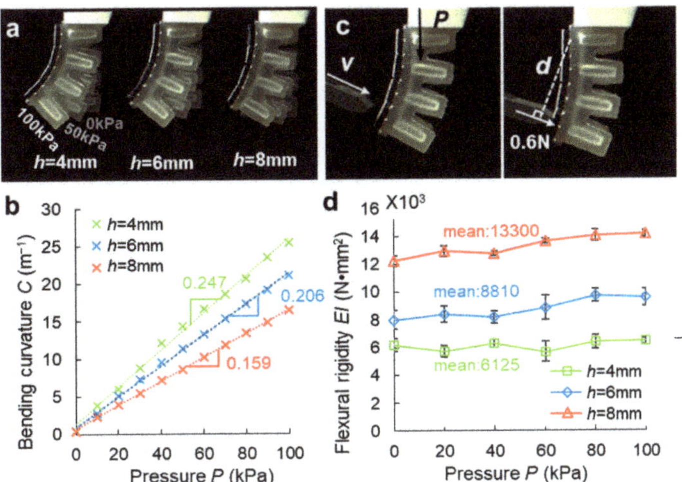

Figure 4. (**a**) The bending state of three SPAAs with rib heights of 4 mm, 6 mm, and 8 mm at 0 kPa, 50 kPa, and 100 kPa, respectively. (**b**) The experimental results of the bending curvature of three SPAAs with different rib heights under pressure drive. (**c**) The test flow of SPAA's flexural rigidity. (**d**) The experimental results of the flexural rigidity of three SPAAs with different rib heights under pressure drive.

4.2. Contact State and Mechanical Properties

Figure 5a presents the flow of the adhesion-peeling performance test of the SPAA. (1) Approach: the SPAA approaches the cylinder fixed vertically at 0.15 mm/s horizontally. (2) Preload: AP1 comes in contact with the cylinder while the SPAA approaches. When the preload reaches 1 N, the SPAA stops moving. (3) Adhesion: the SPAA contacts and adheres to the cylinder gently as the air pressure increases slowly. (4) Peeling: the SPAA peels off from the cylinder at a speed of 0.15 mm/s with release angles θ_p of 90°, 60°, and 30°. (5) Separate: the SPAA stops moving until it separates from the cylinder completely. The left pictures of the five stages are the images captured by the side camera. The four white dotted boxes from top to bottom are the areas of AP1~AP4. The bright spots in dotted boxes represent the contact areas (based on the principle of frustrated total reflection in Section 3.2). The right pictures are the images captured by the front camera.

The bending state, contact area, and adhesion force of the SPAA were recorded simultaneously during the whole test. Figure 5b shows a typical curve of tangential adhesion force (F_t), normal adhesion force (F_n) of the SPAA, and contact areas of four APs (S_{AP1}~S_{AP4}) versus the time. During the preload, only AP1 contacts the cylinder and generates a normal squeezing force of approximately 1 N. During the adhesion, AP2, AP3, and AP4 are in contact with the arc surface orderly as the air pressure increases. Meantime, the SPAA exerts a small tangential adhesion force to the cylinder away from the root. It is mainly because the inextensible layer of the SPAA inevitably generates a slight tensile deformation during the expansion of the SPAA. During the peeling, the SPAA does not peel off the cylinder once, but generates two peeling processes. In the first peeling, the F_t and F_n go up and S_{AP1}~S_{AP3} go down synchronously. When the F_t and F_n reach the peak, they fall back rapidly, S_{AP1}~S_{AP3} drop to zero at the same time. In the second peeling, the F_t and F_n increase again but at a slower rate, together with the slight decrease in the S_{AP4}. When the F_t and F_n reach the peak, F_t, F_n, and S_{AP4} drop to zero at the same time.

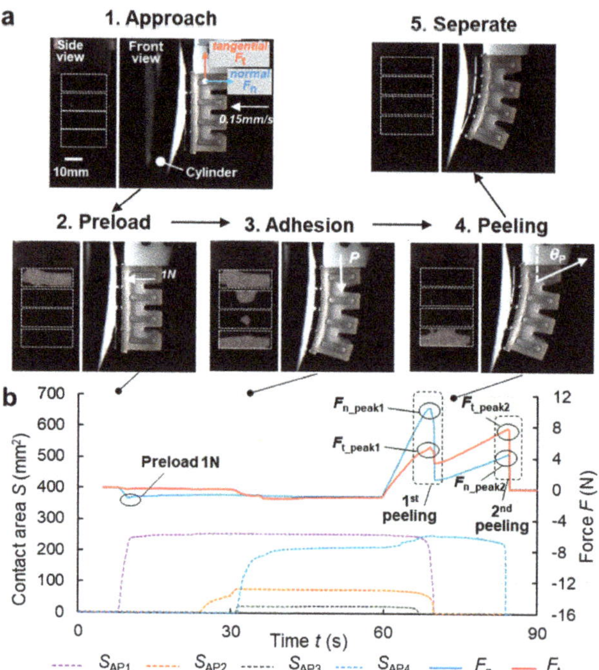

Figure 5. (**a**) The flow of the adhesion-peeling performance test of the SPAA. (**b**) A typical curve of tangential adhesion force (F_t), normal adhesion force (F_n) of the SPAA, and contact areas of four APs (S_{AP1}~S_{AP4}) versus the time.

4.3. Adhesion-Peeling Performance on Large Cylinders

Figure 6 presents the experimental results of the contact area of four APs (S_{AP1}~S_{AP4}) during the adhesion stage. Obviously, the S_{AP4} increases the most with an increase in the rib height. As the rib height increases from 4 to 8 mm, the contact ratio of the end region increases from 40% to 70%, indicating that a higher rib of the SPAA benefits a more sufficient contact. A low standard deviation means that the contact area state of each AP is little affected by the curvature of the cylinder. We also found that AP3 has the lowest contact area with a 15% contact ratio, indicating that the adhesion performance of the area near AP3 is also lower than that of other APs.

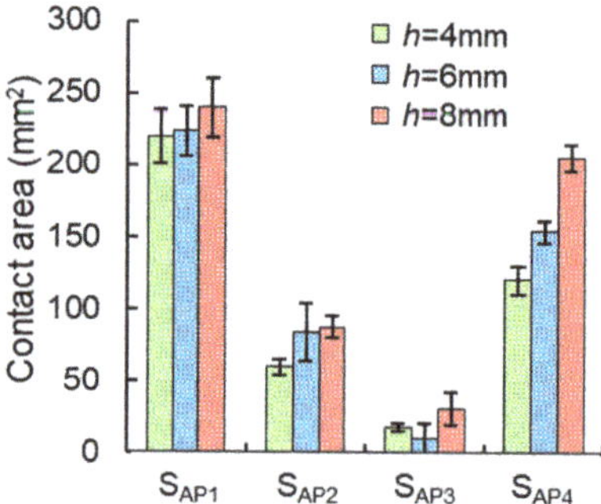

Figure 6. The experimental results of the contact area of four APs (S_{AP1}~S_{AP4}) during the adhesion stage of three SPAAs.

Figure 7 presents the peak forces of F_t and F_n in the first and second peelings of three SPAAs with different rib heights, and defined as F_{t_peak1}, F_{n_peak1}, F_{t_peak2}, and F_{n_peak2}, respectively (Figure 5b). It is evident that with the rib height of the SPAA increasing, the peak adhesive forces show an upward trend. As the rib height increases from 4 mm to 8 mm, the forces F_{n_peak1} and F_{n_peak2} stabilize in the range of 7~10 N and 2~6 N, respectively, and increase slowly. The tangential adhesion force also increases with the rib height increasing, but it is significantly affected by the peeling angle θ_p. Specifically, the force F_{t_peak2} at $\theta_p = 30°$ increases from 6 N to 10~12 N, the most evident increase. The above results mean that the increase in the rib height is conducive to a more outstanding adhesion performance of the SPAA on large cylinders, and the effect is more significant when at a smaller peeling angle. Further, the average increases of F_{n_peak1}, F_{t_peak1}, F_{n_peak2}, and F_{t_peak2} are 20%, 31%, 102%, and 62%, respectively. The increase in the peak adhesion forces in the second peeling is 2~5 times that in the first peeling. Since the F_{t_peak2} and F_{n_peak2} are generated by the peeling off of the AP4 (Section 4.2), it can be verified that the increase in the rib height can significantly improve the adhesion performance of the SPAA's end contact region.

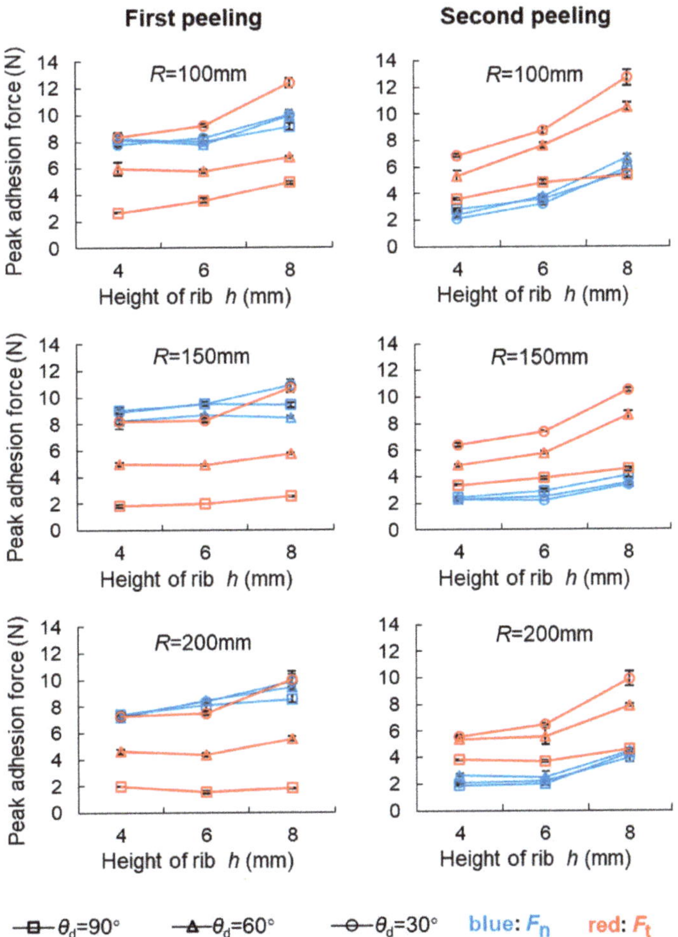

Figure 7. The peak forces of F_t and F_n in the first and second peelings of three SPAAs with different "rib" heights peeling off from cylinders with radii of 100 mm, 150 mm, 200 mm at peeling angles of 90°, 60°, 30°.

The influence of peeling angle θ_p on the peak tangential adhesion force is more significant than that of the peak normal adhesion force. As the θ_p decreases from 90° to 30°, F_{t_peak1} increases from 2~4 N to 8~12 N, with an average increase of 290%; the F_{t_peak2} increases from approximately 4 N to 6~12 N, with an average increase of 95%; while the increase in peak normal adhesion force is very small, only 2%. When peeling at 90°, the F_{t_peak1} is only approximately 10%~30% of the F_{n_peak1}, and the F_{t_peak2} is equivalent to the F_{n_peak2}. However, when peeling at 30°, F_{t_peak1} is basically the same as the F_{n_peak1}, and the F_{t_peak2} is 2~3 times of the F_{n_peak2}. In general, the decrease in θ_p is beneficial to a more excellent tangential adhesion performance of the SPAA on large cylinders, and the effect is more significant when the SPAA has a larger flexural rigidity. In addition, SPAAs with different bending rigidities show a very stable normal adhesion performance when peeling at different angles on large cylinders. Cylinder curvature has little effect on the adhesion performance of the SPAA. As the curvature decreases, the overall adhesion force decreases slightly, indicating that the SPAA can generate a stable adhesion performance when adhering to cylinders with large curvatures.

5. Discussion

The SPAA consists of a top fluidic-driven elastic actuator and four bottom adhesive pads attached with mushroom-shaped microstructured adhesive surfaces. The results of the bending curvature test verify that its controllable and the SPAA can continuously bend under positive air pressure. In comparison with rigid [9,28] or under-actuated [10] soft actuators, the SPAA can actively envelop a wide range of cylinders with different curvatures. Moreover, novel actuating methods such as electrostatic–hydraulic coupled [29,30], chemical reacting [31], magnetic field [32] and biohybrid system actuations [33] have been applied to soft actuators/robots, enabling them to be more flexible, robust and programmable. Combining bio-inspired adhesion with novel actuators is a promising research direction. The flexural rigidity results demonstrate that increasing the rib height can significantly increase the flexural rigidity of SPAA. In comparison to the 10% slight increase in flexural rigidity obtained by reducing the rib weight from 3 to 0.8 mm [21], increasing the rib height from 4 to 8 mm can increase the flexural rigidity of the SPAA by approximately 230%. The results of the adhesion–peeling performance tests indicate that an increase in the rib height and decrease in the peeling angle are conducive to a more outstanding adhesion performance of the SPAA on large cylinders.

5.1. *Comparison between the SPAA and Other Adhesive Units*

Table 1 shows the adhesion capacity of the existing adhesive grippers consisting of soft adhesive actuators. These grippers are all parallel structures that rely on two opposed adhesive actuators to envelop and lift the target. The lifting capacity of the gripper is a combination of two actuators' adhesion capacities. In addition, the two adhesive actuators are mechanically fixed and assembled into an adhesive gripper, so there is a distance between the two actuators. This distance results in an angle between the tangential direction of the actuator and the vertical lifting direction, which is the peeling angle defined in Section 4.2. Furthermore, when the curvature of the gripped surface is constant, the greater the distance, the smaller the peeling angle. Given the significant influence of the peeling angle on the performance of the adhesive actuator (Section 4.3), Table 1 also considers the peeling angle of the adhesive actuator into consideration.

Table 1. Comparison of the adhesion performance of the existing adhesive grippers consisting of soft adhesive actuators.

Soft Adhesive Gripper	Actuating Technology	Single Unit Size (mm)	Detach Force (N)/ Detach Angle (°)/ Radium (mm)
[13]	Fluidic–elastic actuator (rubber)	60 ∗ 20	18/≈60/75 9.5/≈75/101.5
[14]	Fluidic–elastic actuator (fabric)	35 ∗ 24	4.3/90/14
[5]	Under-actuated	≈35 ∗ 100	5.5/≈45/100 12/≈55/150
[34]	Under-actuated	50 ∗ 32	5.5/≈45/100 3.5/≈60/200 1.8/≈80/400
[35]	Shape memory alloy-actuated	100 ∗ 15	10/0/62.5

By comparison, the structural dimensions of adhesive actuators in Table 1 are all in the order of 10 to 100 mm, which are similar to the dimension of SPAA. Most performance tests of the grippers focus on adhering to curved surfaces with a radius of less than 100 mm, and the lifting capacity of grippers decreases as the radius of the curved surface increases, which is similar to the SPAA (except for the gripper in the reference [5], where the adhesion actuator is long, and an increase in the radius of the curved surface increases the adhesion contact area, thereby increasing the adhesive lifting force). Nevertheless, the mean values of two peak normal adhesion forces when the SPAA with a rib height of 8 mm peeling off from a cylinder with a radius equal to 100 mm at a peeling angle of 90° are 9.8 N and 5.4 N, respectively. Both of these two normal adhesion forces of a single SPAA approach or even exceed the lifting force (the combination of two actuators' adhesion forces) of adhesive grippers summarized in Table 1 (except for the grippers in the references [13]). In particular, when the gripper in the reference [13] adheres to a cylinder with a radius of 75 mm, the combined adhesive lifting force of its two adhesive actuators reaches 18 N, which is 1.5~4 times the adhesion performance of a single SPAA. This is mainly because the radius of the gripped cylinder and the peeling angle of the gripper's two actuators are all smaller than those of the SPAA (As mentioned in Section 4.3, a small radius of the cylinder or peeling angle contributes to an excellent adhesion performance). When the SPAA with a rib height of 8 mm adheres to the cylinder with a 100 mm radius and peels off at 60°, the resultant adhesion force reaches 11.8 N. It can be predicted that when two SPAAs are integrated into a parallel gripper to grip a cylinder with a 75 mm radius, their performance will reach or even exceed 18 N. The gripper in the reference [5] achieves an adhesive lifting force of 12 N when adhering to the cylinder with a 150 mm radius, which is slightly higher than the peak resultant adhesion force of 10.3 N in the first peeling and 9.2 N in the second peeling of the SPAA with an 8 mm rib height when adhering to the cylinder with a 150 mm radius and peeling off at 60°, due to the fact that the adhesive actuator of the gripper in the reference [5] has twice the size of the SPAA. However, when evaluated in terms of the adhesion strength (adhesive force per unit area), the normal adhesion strength of the SPAA is approximately three times greater than that of the adhesive actuator of the gripper in the reference [5].

In summary, most of the existing adhesive grippers are used for gripping curved surfaces with a radius of less than 100 mm. The adhesive lifting performance decreases obviously with the radius increase. However, the SPAA shows excellent adhesion performance when adhering to large cylinders, and the normal adhesion performance is little affected by changes in the radius. The magnitude and stability of the SPAA's adhesion performance are better than that of the soft adhesive grippers summarized in Table 1.

5.2. Optimization for High Contact Ratio

Figure 6 shows that the contact ratio between AP2 and AP3 is less than 30%, and Figure 8a shows that the contact area between AP2 and AP3 is mainly in the middle when SPAA contacts the curved surface. It is obvious that the adhesion performance of the region near AP2 and AP3 is not fully utilized.

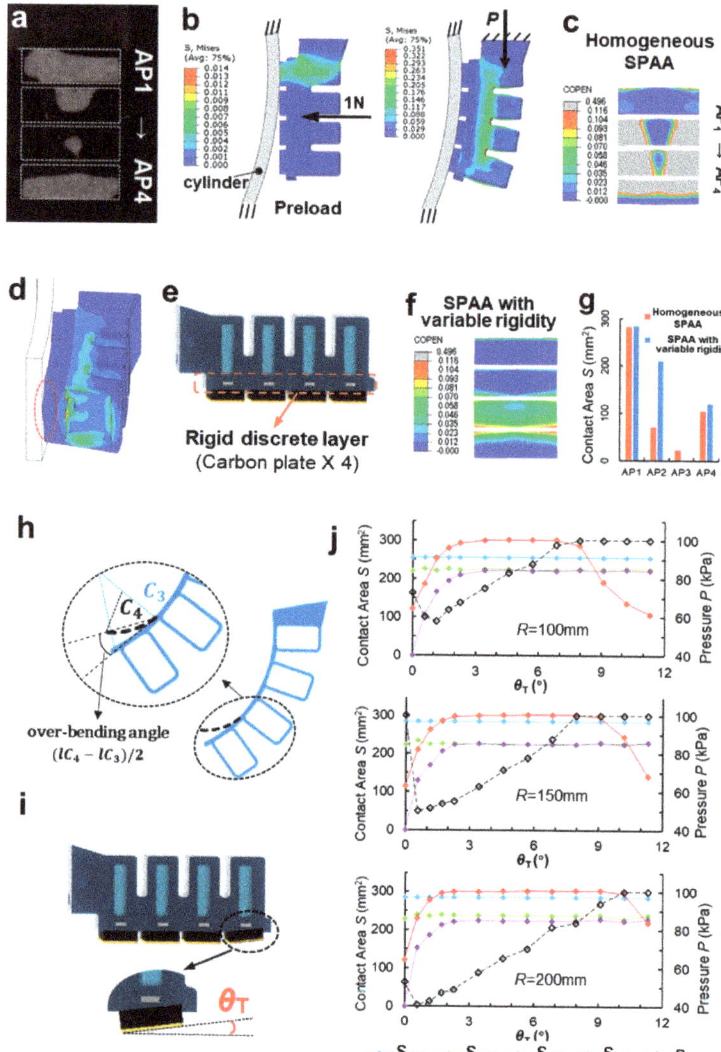

Figure 8. (**a**) The contact state of the SPAA during adhesion. (**b**) The finite element model describing the contact of the SPAA. (**c**) The simulation results of the contact areas of four Aps of the original SPAA with homogeneous structure. (**d**) The simulated cross-section of the SPAA in contact with the cylinder. (**e**) The optimized SPAA with a rigid–flexible coupled hierarchical structure. (**f**) The simulation results of the contact areas of four APs of the optimized SPAA with variable rigidity. (**g**) The increase. (**h**) Illustration of the over-bending angle of PN4 with respect to PN3. (**i**) The optimized SPAA with a differentiated AP structure. (**j**) The contact area and air pressure of each AP versus θ_T (0°~12°) when the SPAA with an 8 mm rib height reaches the steady adhesion.

To investigate the cause of this defect, a FEM (Figure 8b) that describes the contact state of the SPAA was established, where the cylinder was fixed on the bottom side of the PSAA, and the adhesive behavior of the interfacial contact was described by the cohesive model [24,36]. In the preload stage, the PSAA approached and contacted the cylinder at a speed of 0.15 mm/s until the interfacial contact force reached 1 N. In the adhesion stage, the SPAA bent and enveloped the cylinder as the pressure P was applied to the inner wall of SPAA. The simulation results of the contact areas of four Aps (Figure 8c) agree well with the experimental results (Figure 8a). The simulated cross-section of the SPAA in contact with the cylinder is shown in Figure 8d. Obviously, when the SPAA is driven by positive pressure, the inextensible layer also passively expands from its initial plane into an arc, driving Aps to bend radially into an arc. As a result, only small raised areas in the middle of Aps are in contact with the cylinder, although the SPAA has enveloped the cylinder well. We have exploited the inherent flexibility of soft materials to enable better envelope adaptation of the SPAA than conventional rigid actuators, but have neglected that flexibility also reduces the rigidity of the SPAA's bottom side, which affects the contact adaptation. Studies on bio-adhesive systems have shown that a hierarchical structure with variable rigidity facilitates an adequate fit of the gecko toe to the surface [37]. Studies on biomimetic soft adhesive actuators have improved contact adaptation by adjusting the rigidity of the dorsal muscles [13]. In conclusion, the hierarchical structure with rigid–flexible coupled rigidity is beneficial in improving the contact adaptation of the SPAA without affecting its bending performance.

To solve the inadequate contact of the SPAA, we should increase the radial flexural rigidity of the inextensible layer by an order of magnitude without affecting the layer's axial flexural rigidity. Here, we propose a solution of embedding a rigid discrete layer inside the inextensible layer of the SPAA. It is important that the rigid discrete layer should have a high elastic module to resist the bending deformation of the inextensible layer and a low density to minimize the effect of the local mass difference on the performance of the SPAA. Carbon plates have an elastic module of approximately 240 GPa and a density of approximately 1.8 g/cm^3. Compared to ordinary steel, such as aluminum alloy, carbon plates are lighter in weight and higher in bending strength. Finally, four carbon plates (3 mm * 1.5 mm * 30 mm) were selected and embedded inside the inextensible layer (Figure 8e). The optimized SPAA has a hierarchical structure with variable rigidity, which prevents the SPAA's Aps from expanding radially into an arc under positive pressure while maintaining the SPAA's axially flexible bending performance. The FEM results in Figure 8c,f show that the contact area of the AP2 has changed from an intermediate distribution (homogeneous SPAA) to a balanced radial distribution along the AP2 (SPAA with variable rigidity), and the contact area has increased from 70 to 210 mm^2, an increase of 200% (Figure 8g).

The hierarchical structure with variable rigidity helps the APs to make uniform contact with the cylinder along the radial direction. However, the SPAA does not fully contact the cylinder along the axial direction. The results of Figure 8g show that AP3 is barely in contact with the surface, and the contact ratio of AP4 is only about 50%. From Figure 2b, it can be intuitively obtained that the bending curvature of PN4 (C_4) is slightly greater than that of PN3 (C_3), resulting in an over-bending angle of PN4 with respect to PN3, i.e., $(lC_4 - lC_3)/2$ (Figure 8h). This over-bending angle directly causes AP4 to squeeze tightly with the cylinder, while AP3 cannot contact the cylinder.

In order to improve the uniform contact along the radial direction of the SPAA, we propose a differentiated AP structure design by tilting the AP4 to compensate for the axial contact imbalance caused by excessive bending of the PN4, as shown in Figure 8i. To clarify the effect of the tilt angle θ_T of AP4 on the contact state along the SPAA axial direction, the contact area and air pressure of each AP versus θ_T (0°~12°) when the SPAA with an 8 mm rib height reaches the steady adhesion (contact areas of all APs reach the peak and remain stable) on cylinders with radii of 100 mm, 150 mm and 200 mm, respectively, were simulated, and the results are shown in Figure 8j. It is obvious that S_{AP1} and S_{AP2} are almost unaffected by θ_T, while S_{AP3} and S_{AP4} rise rapidly to a peak as θ_T increases from 0° to 3°, after which S_{AP3} is maintained. However, as θ_T continues to raise, S_{AP4} begins to fall, and the pressure P also reaches 100 kPa at this point. Since the excessive θ_T requires higher air pressure to drive larger bending deformation of the SPAA to fit AP4 to the cylinder, the θ_T at the point S_{AP4} begins to fall increases as the radius of the cylinder increases. The results of the SPAA on all three cylinders show that a good contact state can be achieved with θ_T between 3° and 6°, and the contact ratio of four APs is between 70% and 100%, which is a noticeable optimization compared to the SPAA with untitled AP4. Too low a θ_T would result in inadequate contact of AP3, and too high a θ_T would increase the air pressure to reach a steady-adhesion state. The final determination is θ_T = 5°, at which the overall contact ratio of the SPAA reaches over 80%.

6. Conclusions

In this study, we designed a SPAA comprising a top fluidic-driven elastic actuator and four bottom adhesive pads for adhering to large cylinders. The excellent envelope adaptation of the SPAA to cylinders of various radii and enhancement of the adhesion performance of the SPAA by increasing the flexural rigidity were demonstrated via FEA and experiments. To address the low contact ratio of the SPAA, a rigid–flexible coupling hierarchical structure with embedded discrete carbon plates and differentiated AP structures were proposed to optimize the contact state. The finite element simulation results verified that the aforementioned structural optimizations contributed to adequate contact between the SPAA and cylinders along the radial and axial directions, respectively, thereby increasing the overall contact ratio to more than 80%. This study demonstrates the influence of rigidity on the adhesion performance of soft actuators, which is conducive to advancing research on soft adhesive actuators with characteristics such as variable rigidity and hierarchical structure, enhancing the performance of adhesion grippers, and expanding the application areas of adhesion technology.

In future, there are three immediate directions in which this study should be extended. First, the effects of the contact stiffness of the adhesive layer should be considered to improve the contact state of the soft adhesive actuator. Second, engineering the bionic features of hierarchical adhesive structures, reversible adhesion, and intelligent sensing into reality will be necessary to obviously improve the performance of bionic adhesive devices. Third, the application of novel actuating methods, such as electrostatic-hydraulic coupled, chemical reacting, magnetic field, and biohybrid system actuations, to adhesive actuators/grippers should be explored.

Author Contributions: L.W. and Z.W. (Zhouyi Wang) designed the research. L.W. and Q.J. carried out the design and analysis of the structure. Z.W. (Zhiyuan Weng) carried out the fabrication. L.W. and Q.Y. carried out the experiments. All the authors wrote the manuscript. All authors have read and agreed to the published version of the manuscript.

Funding: This work was supported by the National Natural Science Foundation of China (Grant No. 51975283 to Zhouyi Wang) and Basic Research Program of Shenzhen (JCYJ20210324122812033 to Zhouyi Wang).

Data Availability Statement: The data generated and/or analyzed during the current study are not publicly available for legal/ethical reasons but are available from the corresponding author up-on reasonable request.

Acknowledgments: We thank Yuxi Feng, YiPing Feng and Junsheng Yao for assistance with fabrication and data collection. We gratefully acknowledge the helpful comments and suggestions from the anonymous reviewers.

Conflicts of Interest: The authors declare no conflict of interest.

References

1. Laschi, C.; Mazzolai, B.; Cianchetti, M. Soft robotics: Technologies and systems pushing the boundaries of robot abilities. *Sci. Robot.* **2016**, *1*, eaah3690. [CrossRef]
2. Shintake, J.; Cacucciolo, V.; Floreano, D.; Shea, H. Soft robotic grippers. *Adv. Mater.* **2018**, *30*, e1707035. [CrossRef]
3. Hawkes, E.W.; Jiang, H.; Christensen, D.L.; Han, A.K.; Cutkosky, M.R. Grasping without squeezing: Design and modeling of shear-activated grippers. *IEEE Trans. Robot.* **2017**, *34*, 303–316. [CrossRef]
4. Brown, E.; Rodenberg, N.; Amend, J.; Mozeika, A.; Steltz, E.; Zakin, M.R.; Lipson, H.; Jaeger, H.M. Universal robotic gripper based on the jamming of granular material. *Proc. Natl. Acad. Sci. USA* **2010**, *107*, 18809–18814. [CrossRef]
5. Alizadehyazdi, V.; Bonthron, M.; Spenko, M. An electrostatic/gecko-inspired adhesives soft robotic gripper. *IEEE Robot. Autom. Lett.* **2020**, *5*, 4679–4686. [CrossRef]
6. Tao, D.; Gao, X.; Lu, H.; Liu, Z.; Li, Y.; Tong, H.; Pesika, N.; Meng, Y.; Tian, Y. Controllable anisotropic dry adhesion in vacuum: Gecko inspired wedged surface fabricated with ultraprecision diamond cutting. *Adv. Funct. Mater.* **2017**, *27*, 1606576. [CrossRef]
7. Modabberifar, M.; Spenko, M. A shape memory alloy-actuated gecko-inspired robotic gripper. *Sens. Actuators A Phys.* **2018**, *276*, 76–82. [CrossRef]
8. Modabberifar, M.; Spenko, M. Development of a gecko-like robotic gripper using Scott–Russell mechanisms. *Robotica* **2020**, *38*, 541–549. [CrossRef]
9. Dadkhah, M.; Zhao, Z.; Wettels, N.; Spenko, M. A self-aligning gripper using an electrostatic/gecko-like adhesive. In Proceedings of the 2016 IEEE/RSJ International Conference on Intelligent Robots and Systems (IROS), Daejeon, Korea, 9–14 October 2016.
10. Ruotolo, W.; Brouwer, D.; Cutkosky, M.R. From grasping to manipulation with gecko-inspired adhesives on a multifinger gripper. *Sci. Robot.* **2021**, *6*, eabi9773. [CrossRef]
11. Hashizume, J.; Huh, T.M.; Suresh, S.A.; Cutkosky, M.R. Capacitive sensing for a gripper with gecko-inspired adhesive film. *IEEE Robot. Autom. Lett.* **2019**, *4*, 677–683. [CrossRef]
12. Hawkes, E.W.; Christensen, D.L.; Han, A.K.; Jiang, H.; Cutkosky, M.R. Grasping without squeezing: Shear adhesion gripper with fibrillar thin film. In Proceedings of the 2015 IEEE International Conference on Robotics and Automation (ICRA), Seattle, WA, USA, 26–30 May 2015.
13. Glick, P.; Suresh, S.A.; Ruffatto, D.; Cutkosky, M.; Tolley, M.T.; Parness, A. A soft robotic gripper with gecko-inspired adhesive. *IEEE Robot. Autom. Lett.* **2018**, *3*, 903–910. [CrossRef]
14. Hoang, T.T.; Quek, J.J.S.; Thai, M.T.; Phan, P.T.; Lovell, N.H.; Do, T.N. Soft robotic fabric gripper with gecko adhesion and variable stiffness. *Sens. Actuators A Phys.* **2021**, *323*, 112673. [CrossRef]
15. Li, L.; Liu, Z.; Zhou, M.; Li, X.; Meng, Y.; Tian, Y. Structures. Flexible adhesion control by modulating backing stiffness based on jamming of granular materials. *Smart Mater. Struct.* **2019**, *28*, 115023. [CrossRef]
16. Song, S.; Sitti, M. Soft grippers using micro-fibrillar adhesives for transfer printing. *Adv. Mater.* **2014**, *26*, 4901–4906. [CrossRef]
17. Song, S.; Majidi, C.; Sitti, M. Geckogripper: A soft, inflatable robotic gripper using gecko-inspired elastomer micro-fiber adhesives. In Proceedings of the 2014 IEEE/RSJ International Conference on Intelligent Robots and Systems, Chicago, IL, USA, 14–18 September 2014.
18. Minsky, H.K.; Turner, K.T. Achieving enhanced and tunable adhesion via composite posts. *Appl. Phys. Lett.* **2015**, *106*, 201604. [CrossRef]
19. Jiang, Q.; Wang, L.; Weng, Z.; Wang, Z.; Dai, Z.; Chen, W. Effect of the structural characteristics on attachment-detachment mechanics of a rigid-flexible coupling adhesive unit. *Biomimetics* **2022**, *7*, 119. [CrossRef]
20. Hao, Y.; Gong, Z.; Xie, Z.; Guan, S.; Yang, X.; Wang, T.; Wen, L. A soft bionic gripper with variable effective length. *J. Bionic Eng.* **2018**, *15*, 220–235. [CrossRef]
21. Alici, G.; Canty, T.; Mutlu, R.; Hu, W.; Sencadas, V. Modeling and experimental evaluation of bending behavior of soft pneumatic actuators made of discrete actuation chambers. *Soft Robot.* **2018**, *5*, 24–35. [CrossRef]
22. Mosadegh, B.; Polygerinos, P.; Keplinger, C.; Wennstedt, S.; Shepherd, R.F.; Gupta, U.; Shim, J.; Bertoldi, K.; Walsh, C.J.; Whitesides, G.M. Pneumatic Networks for Soft Robotics that Actuate Rapidly. *Adv. Funct. Mater.* **2014**, *24*, 2163–2170. [CrossRef]
23. Carbone, G.; Pierro, E.; Gorb, S.N. Origin of the superior adhesive performance of mushroom-shaped microstructured surfaces. *Soft Matter* **2011**, *7*, 5545–5552. [CrossRef]
24. Zhang, L.; Wang, L.; Weng, Z.; Yuan, Q.; Ji, K.; Wang, Z. Fabrication of flexible multi-cavity bio-inspired adhesive unit using laminated mold pouring. *Machines* **2022**, *10*, 184. [CrossRef]
25. Wang, H.; Zhao, F.; Hu, W. Numerical simulation of quasi-static compression on a complex rubber foam. *Acta Mech. Solida Sin.* **2017**, *30*, 285–290. [CrossRef]
26. Yap, Y.L.; Sing, S.L.; Yeong, W.Y. A review of 3D printing processes and materials for soft robotics. *Rapid Prototyp. J.* **2020**, *26*, 1345–1361. [CrossRef]
27. Eason, E.V.; Hawkes, E.W.; Windheim, M.; Christensen, D.L.; Libby, T.; Cutkosky, M.R. Biomimetics. Stress distribution and contact area measurements of a gecko toe using a high-resolution tactile sensor. *Bioinspir. Biomim.* **2015**, *10*, 016013. [CrossRef]
28. Chu, Z.; Deng, J.; Su, L.; Cui, J.; Sun, F. A gecko-inspired adhesive robotic end effector for critical-contact manipulation. *Sci. China Inf. Sci.* **2022**, *65*, 182203. [CrossRef]
29. Marchese, A.D.; Katzschmann, R.; Rus, D.L. A recipe for soft fluidic elastomer robots. *Soft Robot.* **2015**, *2*, 7–25. [CrossRef]

30. Acome, E.; Mitchell, S.K.; Morrissey, T.G.; Emmett, M.B.; Benjamin, C.; King, M.; Radakovitz, M.; Keplinger, C. Hydraulically amplified self-healing electrostatic actuators with muscle-like performance. *Science* **2018**, *359*, 61–65. [CrossRef]
31. Wehner, M.; Truby, R.L.; Fitzgerald, D.J.; Mosadegh, B.; Whitesides, G.M.; Lewis, J.A.; Wood, R.J. An integrated design and fabrication strategy for entirely soft, autonomous robots. *Nature* **2016**, *536*, 451–455. [CrossRef]
32. Kim, Y.; Yuk, H.; Zhao, R.; Chester, S.A.; Zhao, X. Printing ferromagnetic domains for untethered fast-transforming soft materials. *Nature* **2018**, *558*, 274–279. [CrossRef]
33. Park, S.-J.; Gazzola, M.; Park, K.S.; Park, S.; Di Santo, V.; Blevins, E.L.; Lind, J.U.; Campbell, P.H.; Dauth, S.; Capulli, A.K.; et al. Phototactic guidance of a tissue-engineered soft-robotic ray. *Science* **2016**, *353*, 158–162. [CrossRef]
34. Hirano, D.; Tanishima, N.; Bylard, A.; Chen, T.G. Underactuated gecko adhesive gripper for simple and versatile grasp. In Proceedings of the 2020 IEEE International Conference on Robotics and Automation (ICRA), Paris, France, 31 May–31 August 2020.
35. Hu, Q.; Dong, E.; Sun, D. Soft gripper design based on the integration of flat dry adhesive, soft actuator, and microspine. *IEEE Trans. Robot.* **2021**, *37*, 1065–1080. [CrossRef]
36. Tvergaard, V.; Hutchinson, J.W. Effect of strain dependent cohesive zone model on predictions of interface crack growth. *J. Phys. Colloq.* **1996**, *6*, C6-165–C6-172. [CrossRef]
37. Tian, Y.; Wan, J.; Pesika, N.; Zhou, M. Bridging nanocontacts to macroscale gecko adhesion by sliding soft lamellar skin supported setal array. *Sci. Rep.* **2013**, *3*, 1382. [CrossRef]

MDPI
St. Alban-Anlage 66
4052 Basel
Switzerland
www.mdpi.com

Actuators Editorial Office
E-mail: actuators@mdpi.com
www.mdpi.com/journal/actuators

www.ingramcontent.com/pod-product-compliance
Lightning Source LLC
LaVergne TN
LVHW070918170726
843515LV00005B/813

* 9 7 8 3 0 3 6 5 9 8 0 8 6 *